U0942631

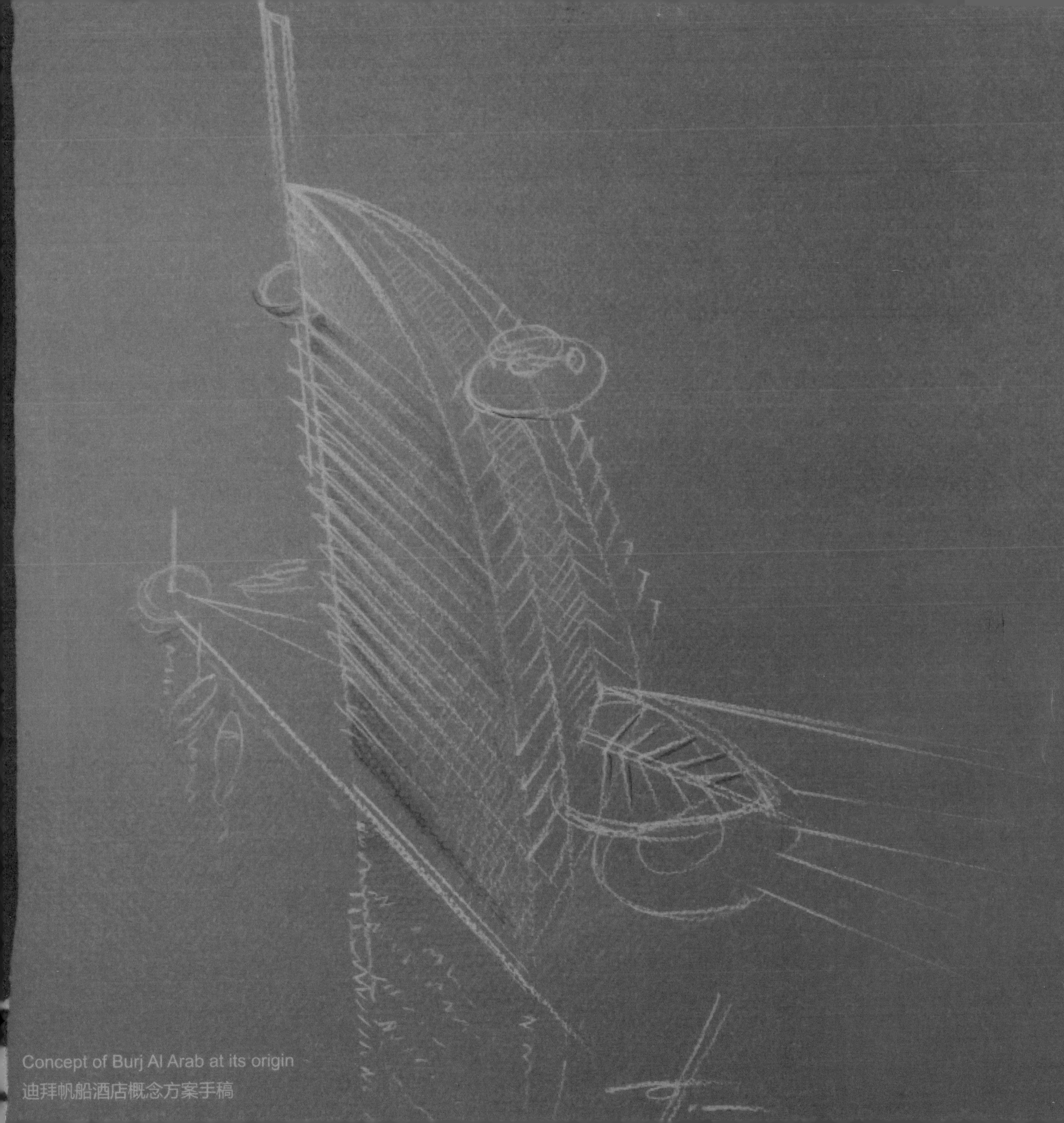

Concept of Burj Al Arab at its origin
迪拜帆船酒店概念方案手稿

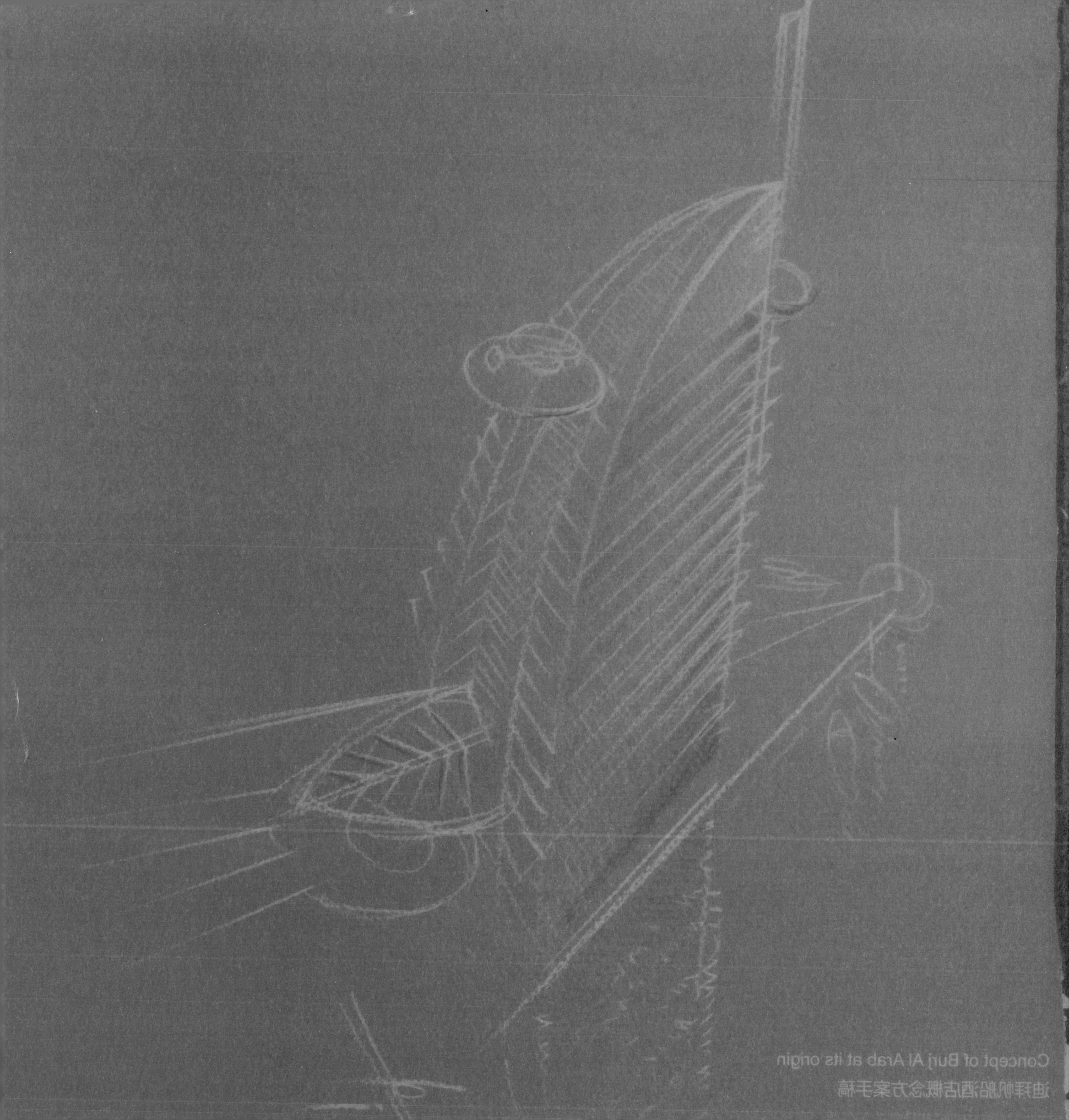

Concept of Burj Al Arab at its origin

迪拜帆船酒店概念方案手稿

CARLOS OTT

卡洛斯・奥特作品集

[加拿大] 卡洛斯・奥特　著

[加拿大] 庄隽　译

著作合同登记号　图字：11-2021-194号

图书在版编目（CIP）数据

卡洛斯·奥特作品集 /（加）卡洛斯·奥特（Carlos Ott）著；（加）庄隽（June Chuang）译. — 杭州：浙江科学技术出版社，2021.9

ISBN 978-7-5341-9841-0

Ⅰ.①卡…　Ⅱ.①卡…　②庄…　Ⅲ.①建筑设计-作品集-加拿大-现代　Ⅳ.①TU206

中国版本图书馆CIP数据核字（2021）第180393号

CARLOS OTT

卡洛斯·奥特作品集

［加拿大］卡洛斯·奥特　著　［加拿大］庄隽　译

出版发行　**浙江科学技术出版社**

杭州市体育场路347号　邮政编码：310006

办公室电话：0571-85176593

销售部电话：0571-85176040

网址：www.zkpress.com

E-mail：zkpress@zkpress.com

排　　版　杭州兴邦电子印务有限公司

印　　刷　浙江海虹彩色印务有限公司

开　本	889 × 1194　1/12	**印　张**	28 1/3
字　数	280 000	**插　页**	1
版　次	2021年9月第1版	**印　次**	2021年9月第1次印刷
书　号	ISBN 978-7-5341-9841-0	**定　价**	498.00元

责任编辑　卢晓梅　　**责任校对**　张　宁

责任美编　曹莼君　　**责任印务**　叶文炀

Carlos Ott B.Arch；M. Arch/UD;OAA RAIC

卡洛斯•奥特（建筑学学士；建筑学和城市设计硕士；安大略建筑师学会、加拿大皇家建筑师学会会员）

Academic / 教育背景

Master Degree in Architecture and Urban Design
Washington University, School of Architecture, St. Louis, Missouri, USA, 1972
建筑学和城市设计硕士学位
华盛顿大学建筑学院，美国密苏里州圣·路易斯，1972

Master in Architecture and Urban Design (Fulbright Scholarship)
Hawaii University, Honolulu, USA, 1972
建筑学和城市设计硕士学位 (富布赖特奖学金)
夏威夷大学，美国夏威夷州火努鲁鲁，1972

Bachelor Degree in Architecture
Universidad de Uruguay, Montevideo, Uruguay, 1971
建筑学学士
乌拉圭大学，乌拉圭蒙得维的亚，1971

Architectural Studies
Uruguayan Institute of Preparatory Studies，Montevideo, Uruguay, 1964
建筑学
乌拉圭建筑研究预备学院，乌拉圭蒙得维的亚，1964

Professional / 专业经历

National Council for the School of Architecture
Washington University, St. Louis, Missouri, USA, 1972
美国建筑学院理事会，美国密苏里州圣·路易斯，始于1972

Sociedad de Arquitectos del Uruguay, Montevideo, 1971
乌拉圭建筑师协会，乌拉圭蒙得维的亚，始于1971

Royal Architectural Institute of Canada, Canada, 1977
加拿大皇家建筑师学会，加拿大，始于1977

Ontario Association of Architects, Canada, 1977
安大略建筑师学会，加拿大，始于1977

Toronto Society of Architects, Canada, 1977
多伦多建筑师学会，加拿大，始于1977

Institut Francais d´Architecture, Paris, France, 1983
法国建筑学院，法国巴黎，始于1983

本书参编人员

（按姓氏首字母排序）

GROUP OF EDITORS AND RESEARCHERS

(Alphabetized by Last Name)

ALEJANDRO ACOSTA 亚历桑德罗·阿科斯塔
FERNANDO ALBA 费尔南多·阿尔巴
LUCIA BENAVIDEZ 露西娅·贝纳维德兹
PATRICIA BOGGIO 帕特里夏·博希奥
ALLA CHING SHAN CHEUNG 张静珊
CHRISTOPHER WAI YIU CHOY 蔡伟耀
JUNE CHUANG 庄隽
JIAQIANG HUANG 黄佳强
SILVIA MENENDEZ 西尔维娅·梅内德斯
CARLOS OTT 卡洛斯·奥特
ISABEL OTT 伊莎贝尔·奥特
SANDRA ROSAS 桑德拉·罗萨斯
ZEFENG SHI 施泽峰
XIAOFAN WANG 王宵凡
MENGDIE ZHANG 张梦蝶
YU ZHENG 郑羽
ETHAN YIFAN ZHOU 周意钒
WENJING ZHU 朱文婧

Contents / 目录

L´OPERA BASTILLE
巴士底歌剧院

Cultural / 文化建筑

Paris, Île-de-France, FRANCE / 巴黎, 法兰西大岛, 法国

AWARDS / 获奖：1st Prize in International Design Competition / 国际设计竞赛一等奖
LOCATION / 地址：Rue de Lyon 120, Bastille Square, Quinze-Vingts
CLIENT / 业主：Establishment Public Opera Bastille
AREA / 面积：150 000 m^2
HEIGHT / 高度：50 m
DATE / 设计时间：1984
STATUS / 状态：Completed 1989 / 1989年建成
DESIGN ARCHITECT / 主创建筑师：Carlos Ott
ARCHITECT OF RECORD / 合作方：Cabinet Saubot-Jullien

Located in the Place de la Bastille where two hundred years ago the French Revolution took place, the two hectares site is crossed by three subway lines and faces the canal of St. Martin. Completed on the 14th of July in 1989 for the 200th anniversary of the French Revolution, the building was inaugurated in the presence of 52 heads of state and hosted by President Francois Mitterrand.

法国巴士底歌剧院位于两百年前爆发法国大革命的巴士底狱广场，占地面积达两万平方米，面对圣马丁运河，地下有三条地铁线。歌剧院于1989年7月14日竣工，作为法国大革命200周年的献礼。落成典礼由时任总统弗朗索瓦·密特朗（Francois Mitterrand）主持，52位国家元首共同出席。

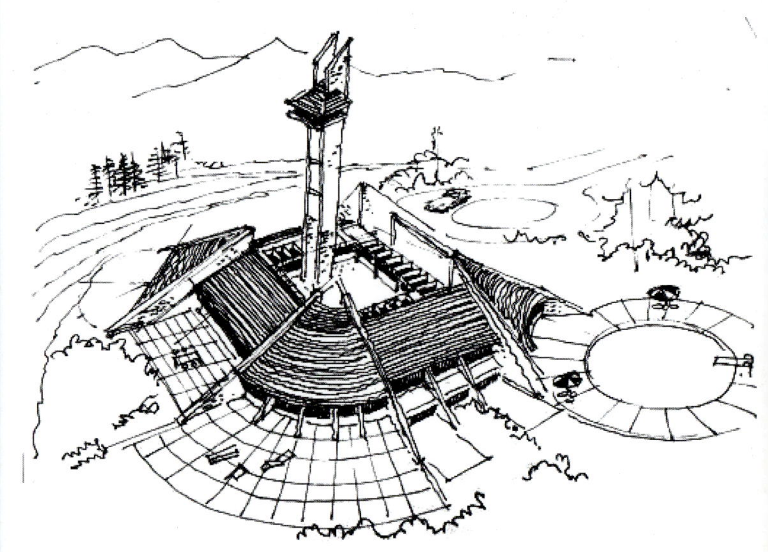

LE CAYROU RESIDENCE
凯鲁山顶私宅

Private Residence / 私人住宅
Auvergne, FRANCE / 奥弗涅, 法国

LOCATION / 地址：Auvergne
CLIENT / 业主：Mr. Michel Vidalenc
AREA / 面积：800 m^2
HEIGHT / 高度：3 storeys / 3层
DATE / 设计时间：1987
STATUS / 状态：Completed 1990 / 1990年建成
DESIGN ARCHITECT / 主创建筑师：Carlos Ott
ARCHITECT OF RECORD / 合作方：Gerard Thiel

Located in a superb site looking down a valley and dam in the centre of the Cantal mountains in Auvergne, in the centre of France; the triangular shaped house takes maximum advantage of the unique sights. A centre courtyard surrounded by corridors provides natural light to the classic automobile collection garage.The steel frame of the roofs and concrete structures are able to respond to the extreme hot and cold weather of the region. Local grey granite clad the exterior walls while the sloping roofs are covered with the typical “lose” stone.

该私人住宅位于法国中部奥弗涅的坎塔尔山脉群山核心区域，这里可以俯瞰山谷和水坝，三角形的房屋设计充分利用了此处独特的自然风光。中央庭院四周走廊环绕，天窗为收藏经典汽车的车库提供了自然光。钢框架屋顶和混凝土结构可有效应对该地区的极端高温和低温气候。外墙覆盖着当地产的灰色花岗岩，屋顶采纳了当地经典的锥形设计，上覆“松散”的石子，造型优雅。

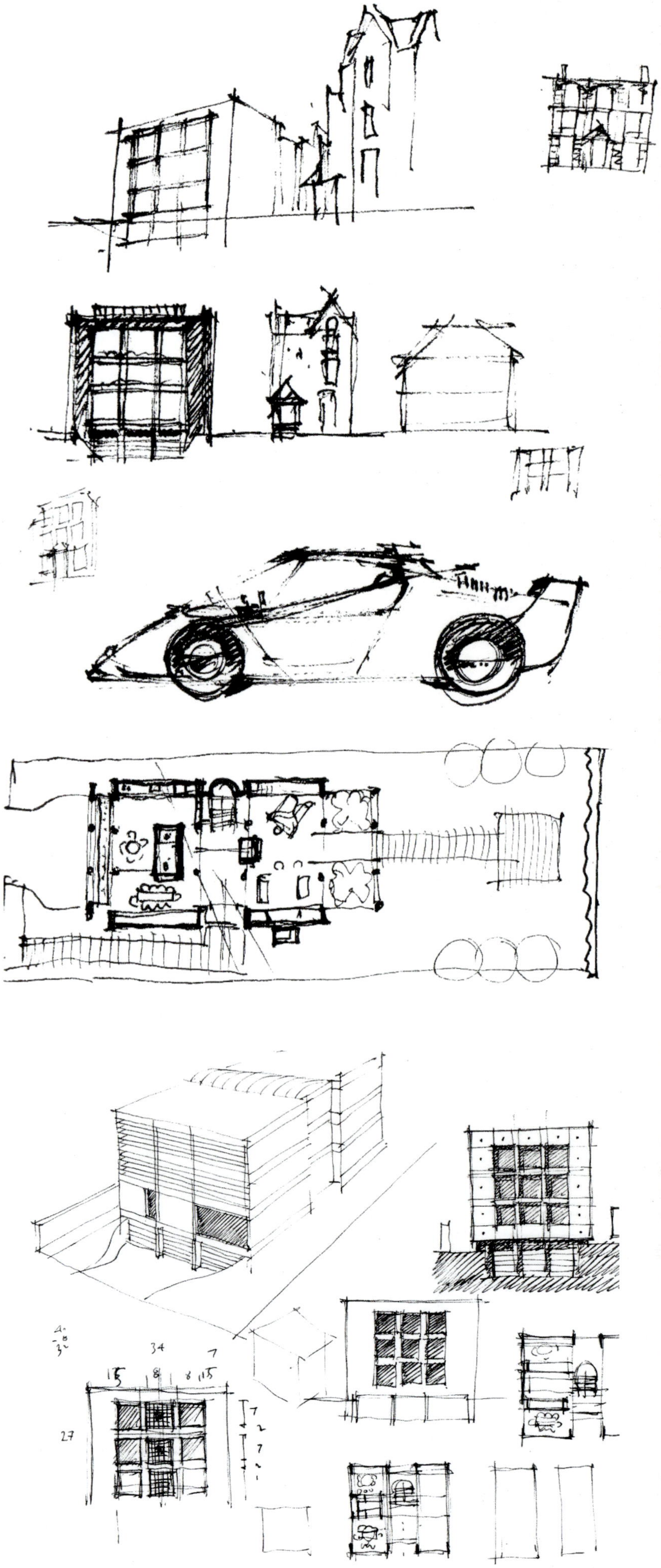

138 LYTTON BOULEVARD RESIDENCE
莱顿大道138号私宅

Private Residence / 私人住宅

Toronto, CANADA / 多伦多, 加拿大

LOCATION / 地址：138 Lytton Boulevard, Toronto
CLIENT / 业主：Carlos Ott
AREA / 面积：600 m^2
HEIGHT / 高度：9 m
DATE / 设计时间：1990
STATUS / 状态：Completed 1993 / 1993年建成
DESIGN ARCHITECT / 主创建筑师：Carlos Ott
ARCHITECT OF RECORD / 合作方：Romano Erba

Situated in a very traditional neighborhood of Toronto, this modern architecture respects the proportions of adjacent houses by utilizing a very contemporary architectural language.

这幢后现代的私宅坐落在多伦多一个非常传统的高端社区，既使用了现代建筑语言，又保持了与相邻房屋的和谐。

UNION NATIONAL BANK HEADQUARTERS
联合国家银行总部

Corporate / 企业总部

Abu Dhabi, UAE / 阿布扎比, 阿联酋

AWARDS / 获奖：1st Prize in International Design Competition / 国际设计竞赛一等奖

LOCATION / 地址：Sheikh Zayed Bin Sultan St. & Al Douj St., Abu Dhabi

CLIENT / 业主：Union National Bank

AREA / 面积：20 000 m^2

HEIGHT / 高度：66 m, 16 storeys / 16层

DATE / 设计时间：1992

STATUS / 状态：Completed 1995 / 1995年建成

DESIGN ARCHITECT / 主创建筑师：Carlos Ott

ARCHITECT / 建筑师：Carlos Ott, NORR

ARCHITECT OF RECORD / 合作方：DUBARCH Architects

The solemn image of the Union National Bank Headquarters achieved by a dark granite clad structure with a matte entrance, leads to the multi-storey banking hall. The top semi-circular volume of the management offices follows the general design of the office floors, assuring strong design cohesion while emphasizing the hierarchy of the upper levels.

联合国家银行总部覆盖着深色花岗岩，巨型入口直达多层的银行营业大厅，一派庄严、肃穆的形象。顶部管理层办公室呈半圆形，遵循了常规办公楼层的设计，既与下面的建筑浑然一体，又强化了高层的威望。

PUNTA SHOPPING
蓬塔购物中心

Commercial Mall / 商业综合体

Punta del Este, Maldonado, URUGUAY / 埃斯特角, 马尔多纳多, 乌拉圭

LOCATION / 地址：Roosevelt Ave. & Los Alpes St.
CLIENT / 业主：Marystay S.A.
AREA / 面积：29 000 m^2
HEIGHT / 高度：3 storeys / 3层
DATE / 设计时间：1992
STATUS / 状态：Completed 1997 / 1997年建成
DESIGN ARCHITECT / 主创建筑师：Carlos Ott
ARCHITECT OF RECORD / 合作方：Raul Wilner

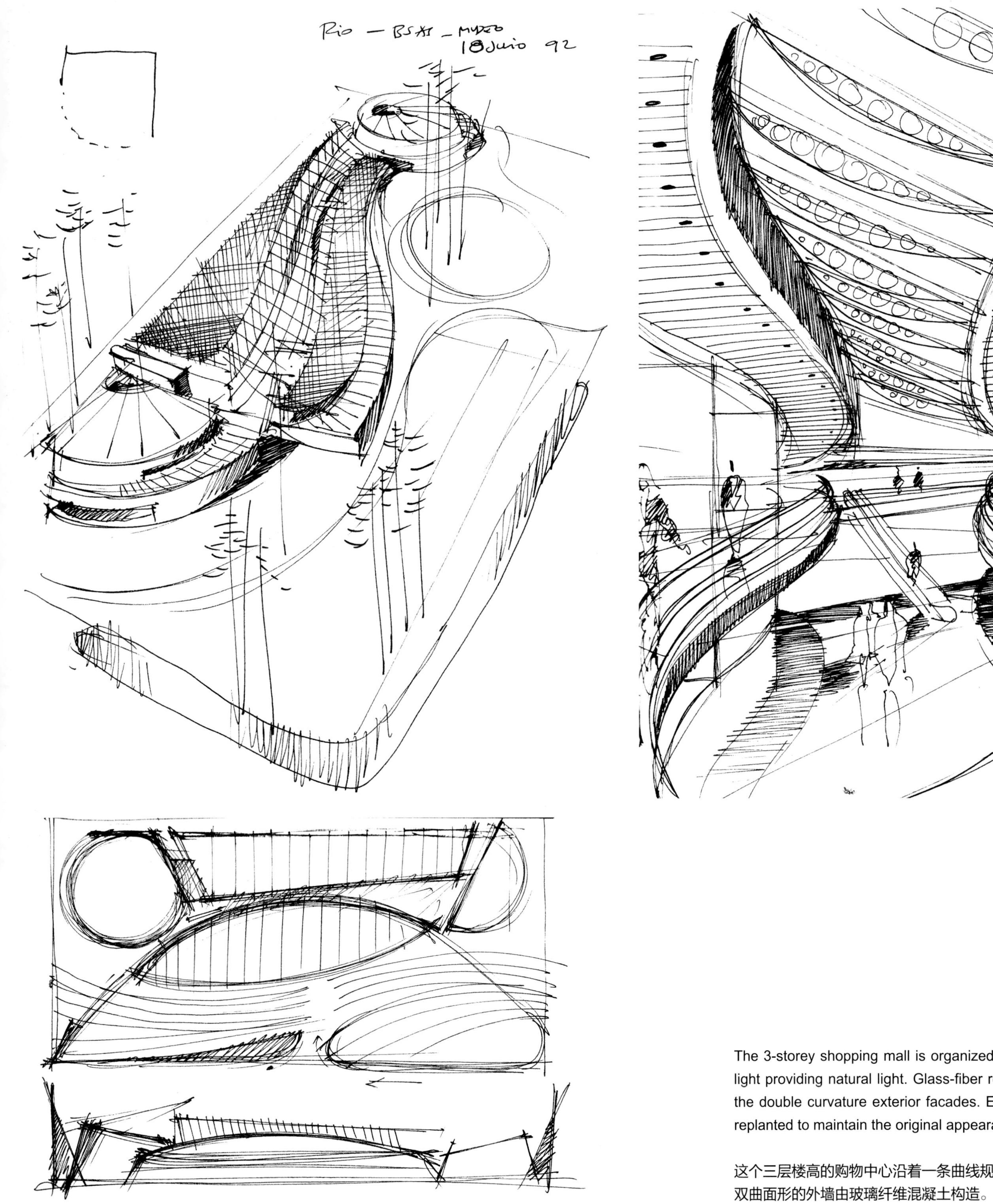

The 3-storey shopping mall is organized along a curvilinear plan with a sky-light providing natural light. Glass-fiber reinforced concrete is used to create the double curvature exterior facades. Existing pine trees are maintained or replanted to maintain the original appearance of the site.

这个三层楼高的购物中心沿着一条曲线规划，天窗为其提供了大量的自然光。双曲面形的外墙由玻璃纤维混凝土构造。选址上原有的松树得以保留或重新移位种植，以最大限度地保持该区域的原生态地貌。

BROADCAST CENTRE DEVELOPMENT SIMCOE PLACE
广播公司总部西姆科大楼

Commercial Offices / 商业办公
Toronto, CANADA / 多伦多, 加拿大

AWARDS / 获奖：1st Prize in International Design Competition / 国际设计一等奖
LOCATION / 地址：Front St. W & Simcoe St., Downtown Toronto
CLIENT / 业主：The Cadillac Fairview Corporation Ltd.
AREA / 面积：125 000 m^2
HEIGHT / 高度：148 m, 33 storeys / 33层
DATE / 设计时间：1992
STATUS / 状态：Completed 1995 / 1995年建成
DESIGN ARCHITECT / 主创建筑师：Carlos Ott
ARCHITECT / 建筑师：Carlos Ott, NORR

The site is adjacent to the Canadian Broadcasting Corporation Headquarters and is across from the Toronto Dome Stadium. Located in one of downtown Toronto main streets, the original project was composed of two towers, only one of which was finally built. Clad in grey and black granite, the tower provides three-star office facilities with large 2000 m^2 office plates with total flexibility for multiple tenants.

西姆科大楼毗邻加拿大广播公司总部，在多伦多圆顶体育场对面。大楼位于多伦多市区主要街道上，原计划由两座塔楼组成，最终只建成了其中一座。该塔楼外墙是灰色和黑色的花岗岩，大楼提供三星级办公设施，每层办公区域达2000 m^2，可灵活容纳多个租户。

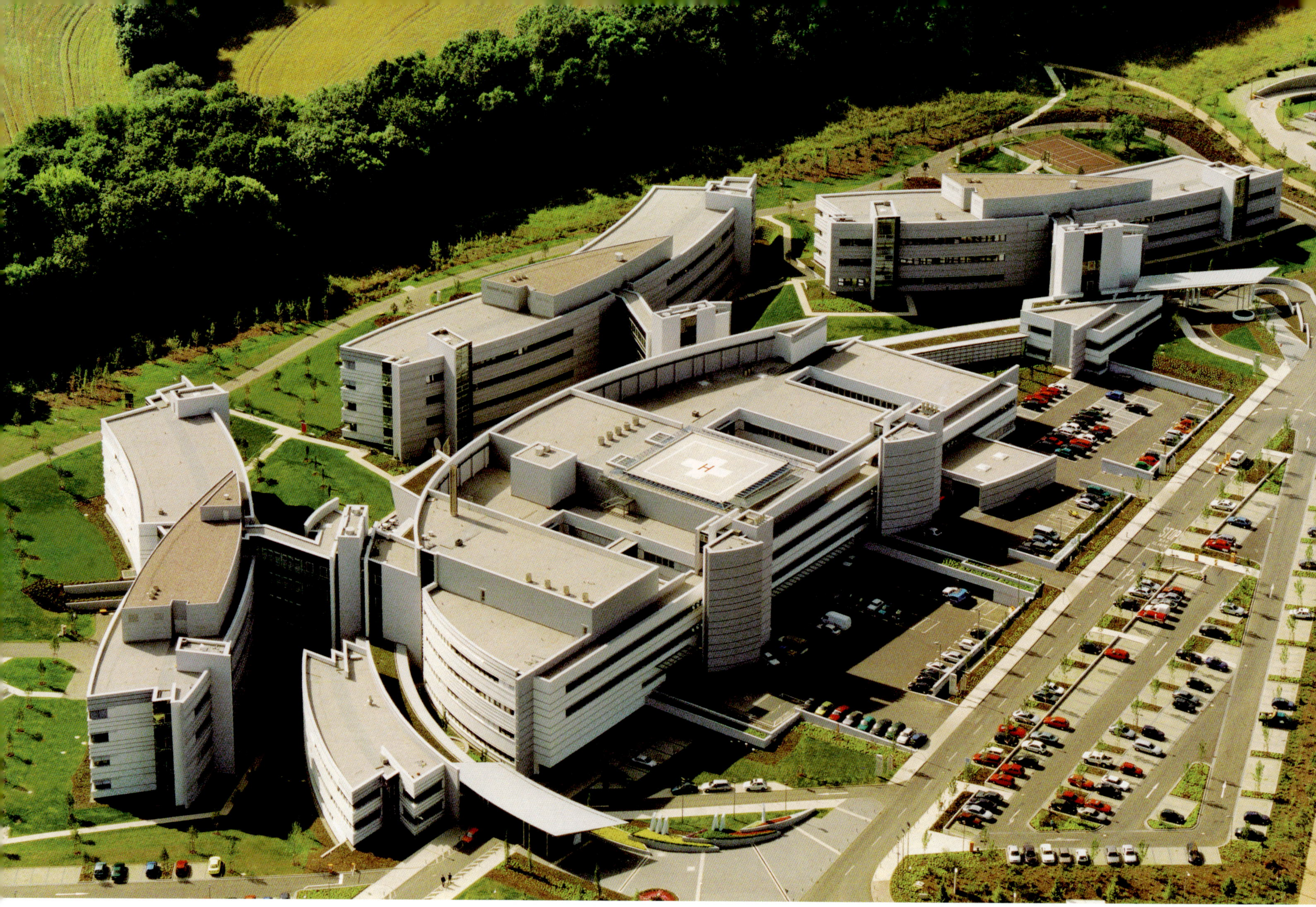

SOPHIEN-UND HUFELAND-KLINIKUM WEIMAR VEREINTE KRANKENHAUS
魏玛医院

Hospital / 医院

Weimar, Thüringia, GERMANY / 魏玛, 图林根, 德国

AWARDS / 获奖：1st Prize in International Design Competition / 国际设计竞赛一等奖
LOCATION / 地址：Henry-van-de-Velde-Straße 2, Weimar
CLIENT / 业主：Ministry of Health of Germany
AREA / 面积：60 000 m^2
HEIGHT / 高度：5 storeys / 5层
DATE / 设计时间：1993
STATUS / 状态：Completed 1998 / 1998年建成
DESIGN ARCHITECT / 主创建筑师：Carlos Ott
ARCHITECT / 建筑师：Plunsgruppe-Ott-WACO
ARCHITECT OF RECORD / 合作方：Alberto Bornes Architect

Situated in the city of Goethe and Schiller, the 600-bed hospital and psychiatric facility and regional pharmacy provides medical services for the residents of the Province of Thüringen. On this long and large site up to 25-metre height drop, this complex is composed of a group of adjacent buildings. The Inpatient Department is next to the main building with a view of the garden. In order to minimize working distance, from its original conception, the use of robots was planned to link different areas. Bridges and elevators are used by patients, visitors, staff and robots alike.

该医院位于歌德和席勒之城魏玛，拥有600张床位、精神病医疗机构和药房，为图林根省的居民提供医疗服务。选址地为长条形，高度落差达25 m，建筑群由大小不同的、相邻的多个单体建筑组合而成，住院部楼宇与主楼相邻，同时可以欣赏公园的风景。为了使员工之间的工作距离最小化，建筑在设计之初便构想了通过使用机器人来连接不同的区域，也设计了很多桥梁和电梯供病人、来访者、医务人员和机器人使用。

MARKS &
ENCER

FOTOUH AL KHAIR CENTRE
FAK 商业中心

Commercial, Residential Complex / 住宅商业综合体
Abu Dhabi, UAE / 阿布扎比, 阿联酋

AWARDS / 获奖：1st Prize in International Design Competition / 国际设计竞赛一等奖
LOCATION / 地址：Airport Rd. & Sheikh Zayed 1st St., Downtown Abu Dhabi
CLIENT / 业主：The Heirs of Sh. Saeed Bin Shakhbout Al Nahyan
AREA / 面积：120 000 m^2
HEIGHT / 高度：87 m, 22 storeys / 22层
UNITS / 户数：224
DATE / 设计时间：1994
STATUS / 状态：Completed 1999 / 1999年建成
DESIGN ARCHITECT / 主创建筑师：Carlos Ott
ARCHITECT / 建筑师：Carlos Ott, NORR
ARCHITECT OF RECORD / 合作方：DEWAN Architects + Engineers

Along airport road, the main thoroughfare linking downtown Abu Dhabi with the international airport, the 3-storey shopping centre roof creates an outdoor plaza for the three residential buildings above it. By organizing the parking lot underground, it makes the entire building face up to the opposite park directly and embraces the scenery.

该商业中心沿着连接阿布扎比市中心和国际机场的主要通道——机场大道，购物中心共有3层楼，楼顶为其上的3座住宅楼创建了一个户外广场。停车场设在地下，使建筑群直面街对面的公园，最大程度利用了这一景观。

AUTOS
OMNIBUS
TAXIS/REMISES

INTERNATIONAL AIRPORT C.C. CARLOS A. CURBELO
埃斯特角国际机场

Airport / 机场

Punta del Este, Maldonado, URUGUAY / 埃斯特角, 马尔多纳多, 乌拉圭

AWARDS / 获奖：1st Prize in International Design Competition / 国际设计竞赛一等奖
LOCATION / 地址：113 km, 93 Route, Laguna del Sauce
CLIENT / 业主：Consorcio Aeropuertos Internacionales S.A.
AREA / 面积：12 000 m^2
HEIGHT / 高度：14 m
DATE / 设计时间：1993
STATUS / 状态：Completed 1997 / 1997年建成
DESIGN ARCHITECT / 主创建筑师：Carlos Ott
ARCHITECT OF RECORD / 合作方：Carlos Ott & Associates

Hertz

Providing access to the Uruguayan Atlantic coast, the airport has been conceived using the shapes of the aircraft design. The steel structure is formed by three parallel forms, one for the vehicular arrival, one for the check in hall, and the last for the gates; the latter two united by a duty free area so that arriving as well as departing passengers will go through it.

作为通往乌拉圭大西洋海岸的空中通道，该机场的造型设计借鉴了飞机的外观。钢结构形成了三个平行空间，一个空间用于车辆到达，一个空间为值机大厅，最后一个空间是机场登机口。后两个空间由一个免税区连接起来，以便到达和离开的旅客都可以通过该区域。

NATIONAL BANK OF ABU DHABI HEADQUARTERS
阿布扎比国家银行总部

Corporate / 企业总部

Abu Dhabi, UAE / 阿布扎比, 阿联酋

AWARDS / 获奖：1st Prize in International Design Competition 1995 / 1995年国际设计竞赛一等奖

LOCATION / 地址：Sheikh Khalif St., Abu Dhabi

CLIENT / 业主：National Bank of Abu Dhabi

AREA / 面积：35 000 m^2

HEIGHT / 高度：173 m, 33 storeys / 33层

DATE / 设计时间：1995

STATUS / 状态：Completed 2002 / 2002年建成

DESIGN ARCHITECT / 主创建筑师：Adel Al Mojil, Carlos Ott Architects

ARCHITECT OF RECORD / 合作方：Architecture & Planning Group

In the centre of downtown Abu Dhabi, the headquarters for the National Bank of Abu Dhabi are designed with two interlocking triangular based volumes, one in glass with the open office space, one clad in blue granite with housing services, elevators and stairs. The bank directory is located under a glass triangular based pyramid at the top of the structure, while the banking hall is at street level under an upside down glass pyramid.

阿布扎比国家银行的总部大楼位于阿布扎比市中心的核心地区，大楼由两个互相连接的、虚实对应的三角体组成：一个是玻璃质的，为开放式办公空间；一个覆盖蓝色花岗岩，为物业服务用房及电梯和楼梯。银行主管层位于大楼顶部，在金字塔形的玻璃结构里，而银行营业大厅位于底层，在倒置的金字塔形玻璃结构里。

ETISALAT HEADQUARTERS
阿联酋电信公司总部

Corporate / 企业总部

Abu Dhabi, UAE / 阿布扎比, 阿联酋

AWARDS / 获奖：1st Prize in International Design Competition / 国际设计竞赛一等奖

LOCATION / 地址：Airport Rd., Abu Dhabi

CLIENT / 业主：Etisalat-Emirates Communications Corporation

AREA / 面积：60 000 m^2

HEIGHT / 高度：26 storeys / 26层

DATE / 设计时间：1995

STATUS / 状态：Completed 2001 / 2001年建成

DESIGN ARCHITECT / 主创建筑师：Adel Al Mojil, Carlos Ott Architects

ARCHITECT OF RECORD / 合作方：Shankland Cox Ltd.

The headquarters of Etisalat, the United Arab Emirates Telecommunications Company, is situated on a major corner site on Airport Road linking Abu Dhabi downtown and the international airport. The tower looks like a cylinder, and a quarter of this cylindrical volume clad in green glass is designed as the open office while a triangular based volume clad in pink granite hides the services and vertical circulation. A major multi-storey hall at street level provides clients reception area.

阿联酋电信公司的总部大楼位于连接阿布扎比市中心和国际机场的机场路的一个主要十字路口。该建筑形似圆柱体，四分之一圆柱的部分采用绿色玻璃幕墙，为开放式办公室，粉红色花岗岩覆盖的三角形方柱状部分隐藏着服务用房和垂直电梯。街道层的多层大厅为客户接待区。

NATIONAL BANK OF DUBAI
迪拜国家银行

Corporate / 企业总部

Dubai, UAE / 迪拜, 阿联酋

AWARDS / 获奖：1st Prize in International Design Competition / 国际设计竞赛一等奖
LOCATION / 地址：Rigga Al Buteen, Dubai
CLIENT / 业主：National Bank of Dubai
AREA / 面积：35 300 m^2
HEIGHT / 高度：125 m, 20 storeys / 20层
DATE / 设计时间：1996
STATUS / 状态：Completed 1998 / 1998年建成
DESIGN ARCHITECT / 主创建筑师：Carlos Ott, NORR
ARCHITECT OF RECORD / 合作方：Dubarch Architects

It is located in the most prominent location on the Dubai Creek, where the original bank was established to export oysters to the Indian subcontinent. The shape of the tower is what the creator of the bank, Mr. Sultan Al Owais, expected, which was designed to be an icon in the Emirates. Inspired by the shape of the dhows, the wooden sail boats that transferred oysters to the adjacent countries and returning with spices and goods. The tower shaped like a boat has a large golden facade appeared as a sail which reflects the light of sunset. It is held by two vertical "masts" clad in light grey granite where circulation and services hide.

迪拜国家银行位于迪拜河最显眼的位置，这里是银行最初成立的地址，选在这里的目的是方便将牡蛎出口到印度次大陆。迪拜银行是其创建者苏尔坦·阿尔·奥维斯（Sultan Al Owais）先生想象中的形态；它的设计旨在成为阿联酋的标志。迪拜河中的单桅帆船给了建筑设计师以启发，木制帆船将牡蛎运到邻国，换回香料和商品。建筑物像船一样，有着巨大的金色玻璃幕墙，呈帆状，随着日落一缕金线自上至下滚动。这金帆由两侧浅灰色花岗岩外立面的"桅杆"支起，其中分布着垂直电梯和服务设施。

Emirates NBD

ABU DHABI CHAMBER OF COMMERCE & INDUSTRY HEADQUARTERS
阿布扎比工商会总部

Corporate / 企业总部

Abu Dhabi, UAE / 阿布扎比, 阿联酋

AWARDS / 获奖：1st Prize in International Design Competition 1997 / 1997年国际设计竞赛一等奖
LOCATION / 地址：Corniche St., Abu Dhabi
CLIENT / 业主：Abu Dhabi Chamber of Commerce and Industry
AREA / 面积：30 000 m²
HEIGHT / 高度：190 m, 20 & 32 storeys / 20层，32层
DATE / 设计时间：1996
STATUS / 状态：Completed 2002 / 2002年建成
DESIGN ARCHITECT / 主创建筑师：Adel Al Moji, Carlos Ott Architects
ARCHITECT OF RECORD / 合作方：Al Mojil Consulting Engineering

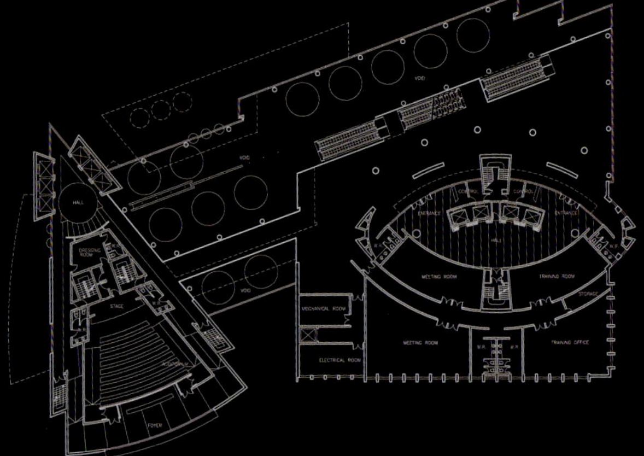

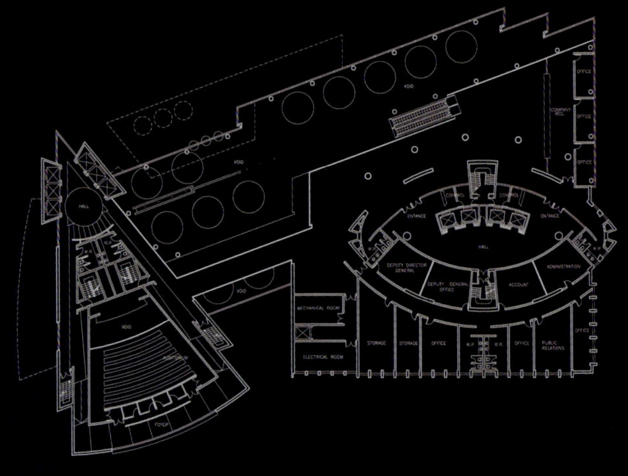

The Abu Dhabi Chamber of Commerce and Industry Headquarters consists of an existing 20-storey office tower and a new 32-storey tower linked by a high-rise glazed bridge, rising from a common podium.

阿布扎比工商会总部大楼由一座现有的20层办公楼和一座新的32层办公楼组成，这两座大楼由一座裙房上的高层玻璃桥连接起来。

ABU DHABI NATIONAL OIL COMPANY HEADQUARTERS
阿布扎比国家石油公司总部

Commercial / 商业

Abu Dhabi, UAE / 阿布扎比, 阿联酋

AWARDS / 获奖：1st Prize in International Competition / 国际设计竞赛一等奖

LOCATION / 地址：Corniche Rd., Abu Dhabi

CLIENT / 业主：Abu Dhabi National Oil Company

AREA / 面积：60 000 m^2

HEIGHT / 高度：38 storeys / 38层

DATE / 设计时间：1996

STATUS / 状态：Completed 2018 / 2018年建成

CONCEPTUAL DESIGN / 概念设计：Adel Al Mojil, Carlos Ott Architects

ARCHITECT OF RECORD / 合作方：Al Mojil Consulting Engineering

Inscribed in a circular layout, a rectangular arch embraces a suspended glazed volume overlooking the nearby sea.

坐落于圆形基地平面之上的阿布扎比国家石油公司总部，外形像方形拱门中悬吊着一个熠熠发光的玻璃体，俯瞰着邻近的大海。

MALVINAS ARGENTINAS INTERNATIONAL AIRPORT USHUAIA
乌斯怀亚国际机场

Airport / 机场

Ushuaia, Tierra del Fuego, ARGENTINA / 乌斯怀亚, 火地岛, 阿根廷

AWARDS / 获奖：1st Prize in International Design Competition / 国际设计竞赛一等奖
LOCATION / 地址：Ushuaia
CLIENT / 业主：London Supply S.A.
AREA / 面积：5700 m^2
HEIGHT / 高度：20 m
DATE / 设计时间：1996
STATUS / 状态: Completed 1997 / 1997年建成
DESIGN ARCHITECT / 主创建筑师：Carlos Ott
ARCHITECT OF RECORD / 合作方：Juan Carlos Sabaté

Located in the most southern city of the world, providing access to Antarctica and in a site along the Beagle Canal that links the Atlantic and Pacific Oceans. The International Airport of Tierra del Fuego was built with glue laminated wood structure due to the short time available to build in a very cold and windy climate. Sharp sloping roofs in steel are conceived to minimize snow loads. Natural stones from the region clad both exterior and interior walls in order to provide a "warm" feeling.

乌斯怀亚国际机场位于世界最南端的城市，可通往南极洲和比格运河沿岸的一个站点，该站点连接大西洋和太平洋。机场采用胶合层压木（工程木）结构建造，以便在寒冷多风的气候下缩短建造时间。倾斜的钢结构屋顶，可最大限度地减少雪荷载可能带来的危险。机场的内墙和外墙均采自本地的天然石材，提供了一种"温暖""熟识"的感觉。

BUSINESS AVENUE
商业大道大厦

Commercial / 商业
Dubai, UAE / 迪拜, 阿联酋

LOCATION / 地址：Sheikh Rashid Rd., Port Saeed, Deira
CLIENT / 业主：German Cars Corporation
AREA / 面积：38 000 m^2
HEIGHT / 高度：11 storeys / 11层
DATE / 设计时间：1996
STATUS / 状态：Completed 2003 / 2003年建成
DESIGN ARCHITECT / 主创建筑师：Adel Al Mojil, Carlos Ott Architects

The golden facade of this 11-storey office building leans towards the end of the runways of the Dubai International Airport. The “U” shaped layout embraces a multi-storey height atrium with views of the exterior and interior landscaped ground floor courtyards.

这座办公大楼有11层楼，大楼的金色外墙朝向迪拜国际机场跑道。大楼的平面布局为“U”形，具有一个多层高的中庭，从大楼里可以欣赏室内外地面庭院的景致。

TELECOMMUNICATIONS TOWER COMPLEX
电信办公综合体

Office / 企业总部

Montevideo, URUGUAY / 蒙得维的亚, 乌拉圭

LOCATION / 地址：Rbla Sudamerica & Guatemala St., Aguada
CLIENT / 业主：Administrative Headquarters of National Telecommunications
AREA / 面积：40 000 m^2
HEIGHT / 高度：164 m, 37 storeys / 37层
DATE / 设计时间：1996
STATUS / 状态：Completed 2002 / 2002年建成
DISIGN ARCHITECT / 主创建筑师：Carlos Ott

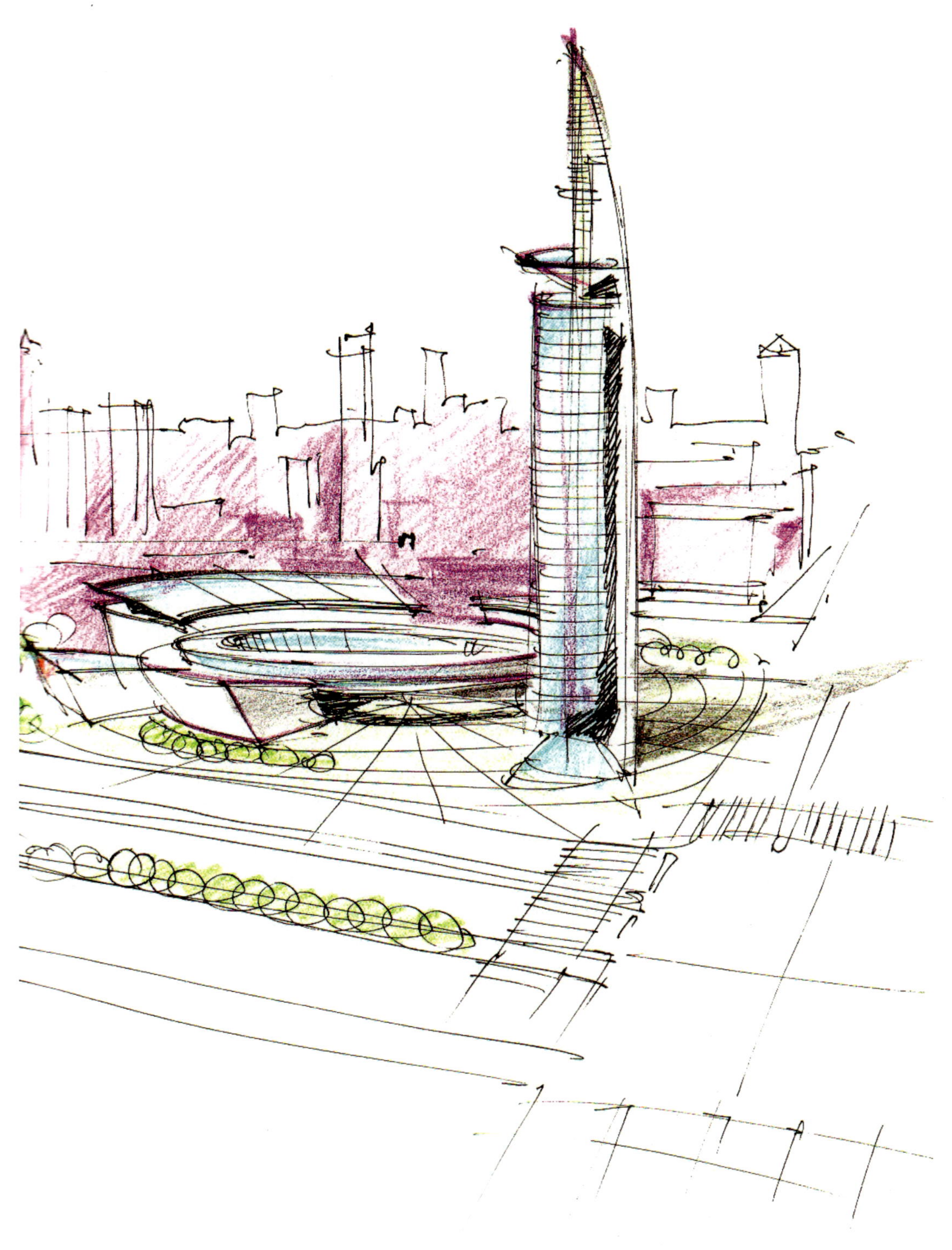

Located in a prominent site, overlooking the bay and Montevideo´s Harbour, this 37-storey iconic building serves as headquarters of Uruguay´s National Telecommunications Company (ANTEL), becoming its symbol. The tower, housing the main offices, with its curved top, delineates the new bold profile in Montevideo´s skyline.

这座高37层的建筑坐落在一个显著的地理位置，从楼上可以俯瞰海湾和蒙得维的亚港口。作为乌拉圭国家电信公司的总部，该建筑已成为其标志。大楼设有总部办公室，曲线形的顶部绘就了蒙得维的亚醒目的新天际线。

PLAZA LIBERTAD
自由广场商业中心

Commercial, Retail / 商业, 零售

Buenos Aires, ARGENTINA / 布宜诺斯艾利斯, 阿根廷

LOCATION / 地址：Cerrito St., & Marcelo T. de Alvear St., Retiro
CLIENT / 业主：Comagasi S.A.
AREA / 面积：9800 m^2
HEIGHT / 高度：46 m, 12 storeys / 12层
UNITS / 户数: 11 open offices / 11间开放办公室
DATE / 设计时间：1997
STATUS / 状态：Completed 2000 / 2000年建成
DESIGN ARCHITECT / 主创建筑师：Carlos Ott
ARCHITECT OF RECORD / 合作方：Estudio Parysow Arquitectos

Located in one of the most important corners on the emblematic artery 9 de Julio of Buenos Aires, in front of Libertad Square, the 12-storey building has a 4 level basement parking served by two car lifts, retail at street level and ten typical office and pent house levels with a concrete structure of slabs without beams with eight columns per floor.

这座12层高的建筑位于布宜诺斯艾利斯标志性核心区9号胡里奥大动脉上最重要的一个拐角处，面对自由广场，设有4层地下停车场，由两部汽车升降机提供服务，零售店铺设在地面层，首层以上是10层标准办公层及顶层阁楼，均设计为无梁混凝土楼板，每层仅设8根柱子。

HILTON DUBAI CREEK
迪拜河希尔顿酒店

Hotel / 酒店

Dubai, UAE / 迪拜, 阿联酋

LOCATION / 地址：Baniyas Rd., Rega Al Buteen
CLIENT / 业主：A. A. Almoosa Enterprises LLC
AREA / 面积：27 000 m^2
HEIGHT / 高度：17 storeys / 17层
UNITS / 户数：150 rooms / 150间客房
DATE / 设计时间：1996
STATUS / 状态：Completed 2001 / 2001年建成
DESIGN ARCHITECT / 主创建筑师：Carlos Ott
ARCHITECT / 建筑师：Adel Al Mojil, Carlos Ott
ARCHITECT OF RECORD / 合作方：Arenco Architectural & Engineering Consultant

موقف خاص لسيارات الأجرة
دبي للمواصلات
Dubai Transport
Private Parking For Taxis
04-2 08 08 08
o Parking

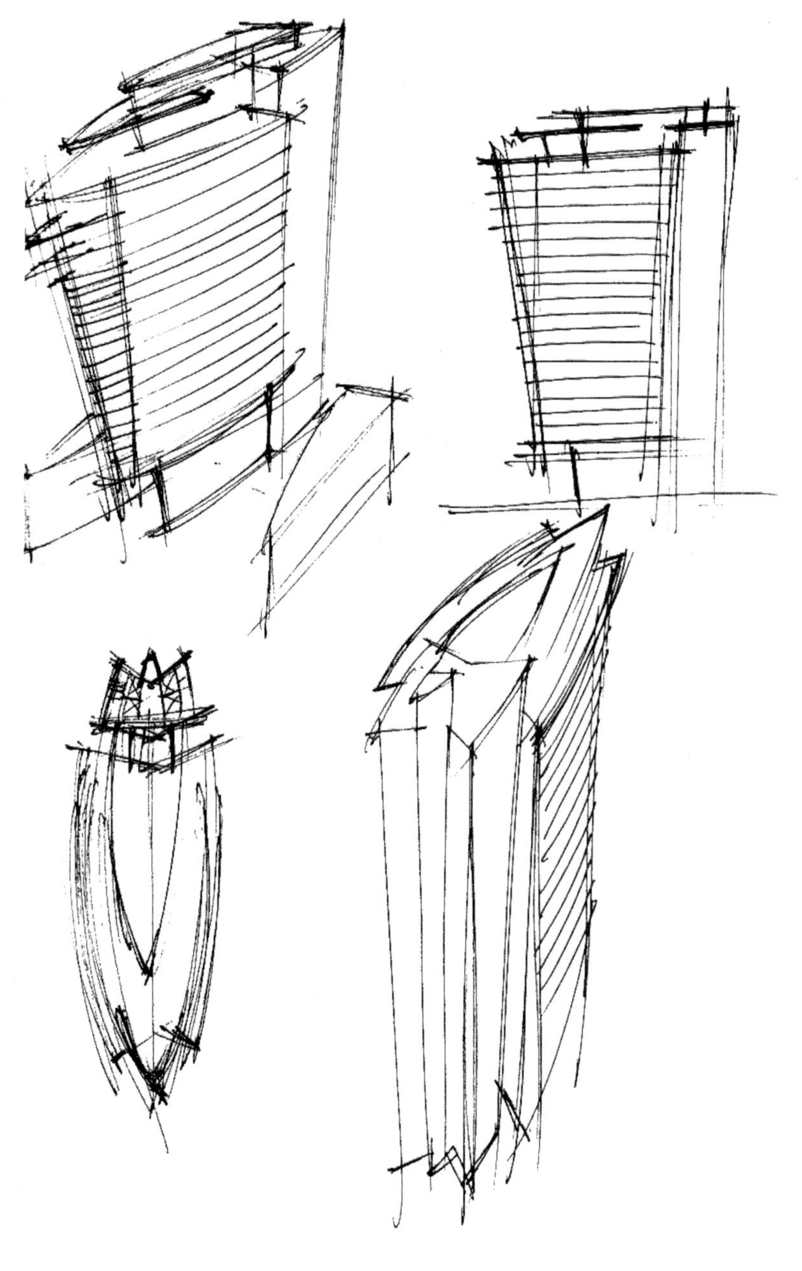

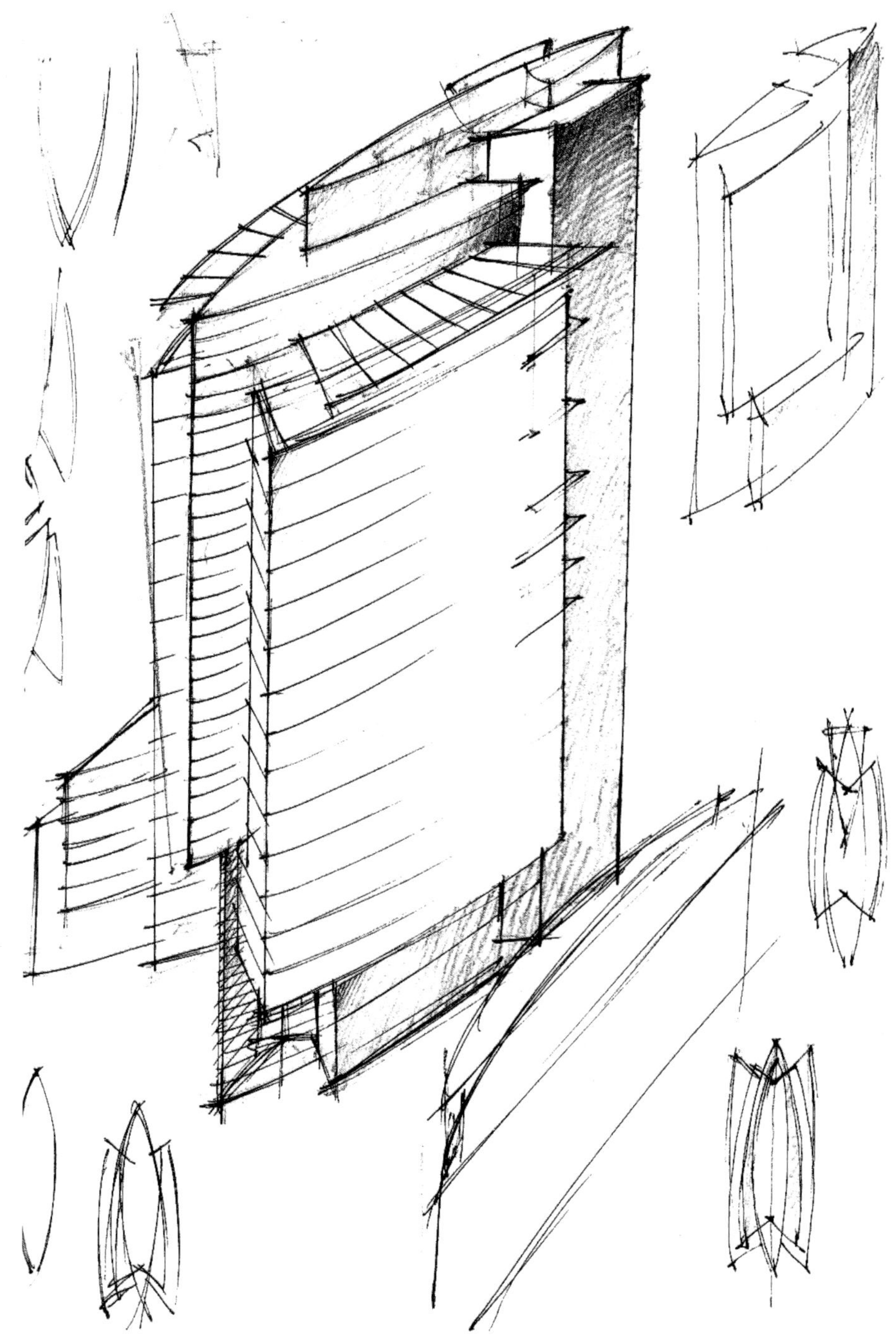

Located in a prime area in Deira, Dubai, with views to the Dubai Creek. The architectural and interior design of the building was done to create a business hotel atmosphere. The black and grey granite of the exterior facades are also found in the interior finishes of lobbies. A major glass stairs links the main two levels around which top restaurants provide different ambiance. The rooms design use cherry wood finishes providing a warm feeling. An outdoor swimming pool is located at the roof level with unique night views of the city.

该酒店位于迪拜德拉的黄金地段，从酒店可以欣赏迪拜河的美景。建筑物的楼宇和室内设计特意营造出商务酒店的氛围。大堂内部装饰使用的黑色和灰色花岗岩呼应了外墙材质。一条主要的玻璃楼梯连接了餐厅的两个楼层，从而提供不一样的氛围。客房设计采用樱桃木饰面，给人温暖的感觉。室外游泳池位于楼顶，住宿客人可在此欣赏城市的独特夜景。

EL CALAFATE INTERNATIONAL AIRPORT CTE. ARMANDO TOLA
埃尔卡拉法特国际机场

Airport / 机场

El Calafate, Santa Cruz, ARGENTINA / 埃尔卡拉法特, 圣克鲁斯, 阿根廷

AWARDS / 获奖：1st Prize in International Design Competition / 国际设计竞赛一等奖
LOCATION / 地址：Ruta Provincial 11 Cal, Lago Argentino
CLIENT / 业主：London Supply S.A.
AREA / 面积：2550 m^2 interior + 750 m^2 partially covered / 室内面积2550 m^2，室外面积750 m^2
HEIGHT / 高度：11 m
DATE / 设计时间：1998
STATUS / 状态：Completed 2000 / 2000年建成
DESIGN ARCHITECT / 主创建筑师：Carlos Ott
ARCHITECT OF RECORD / 合作方：Juan Carlo Sabaté- Alberto Radaelli Arquitectos

The dynamic design of Calafate Airport is a synthesis between high technology equipment and modern materials: large thermal glass surfaces and telescopic access sleeves for aircrafts, with walls made of stone and exposed inclined steel structures supported by rock-clad piers. The angular geometry of the sloping roof covered with double height sheet metal, helps drain heavy snows. The construction parallel to the track, has clear facades opening towards the lake and the mountains, and side facades closed by thick stone-clad masonry walls, are perpendicular to the strong prevailing winds.

埃尔卡拉法特机场的动感设计是高科技设备与现代材料的结晶，例如较大的隔热玻璃表面和飞机的伸缩式检修套筒，以及用石头建成的墙和由石墩支撑着的外露的倾斜钢结构。屋顶的倾斜角几何结构覆盖有双层金属板，有助于积雪的消融。建筑平行于飞机跑道，具有朝向湖泊和山脉的通透立面，山墙由厚厚的石墙组成，并且垂直于盛行风方向。

HANGZHOU GRAND THEATRE
杭州大剧院

Cultural / 文化建筑

Hangzhou, Zhejiang, CHINA / 杭州, 浙江, 中国

AWARDS / 获奖：1st Prize in International Design Competition / 国际设计竞赛一等奖
LOCATION / 地址：Hangzhou CBD, Jianggan District
CLIENT / 业主：Preparation Office of Hangzhou Grand Theatre
AREA / 面积：50 000 m^2
HEIGHT / 高度: 46 m
DATE / 设计时间：1999
STATUS / 状态：Completed 2004 / 2004年建成
DESIGN ARCHITECT / 主创建筑师：Carlos Ott
ARCHITECT / 建筑师：Carlos Ott, PPA Architects
ARCHITECT OF RECORD / 合作方：Hangzhou Architectural Research & Design Institute

The building was the first structure built in the new Hangzhou Central Business District along the Qiantang River.
A master plan had been developed by the municipality to eventually accommodate a convention centre, the new municipality building and a large shopping centre. As the new business area of the city away from the historic West Lake downtown, plots of land were to be offered to potential clients to build new structures around it.
Hangzhou Grand Theatre conceived as a sculpture, is the prime cultural facility of the capital of Zhejiang Province. Known as “a pearl in its oyster´s shell”, the moon-shaped volume formed by complex double curved glass facades and titanium roof, incorporates the exterior green area that surrounds it. The building contains the 1600-seat Opera Hall in the inverted conical shape, besides a 600-seat Concert Hall, a 400-seat Theatre and a rooftop 24-hour open Chinese restaurant with views of the outdoor amphitheatre and the river. Flexible floors, walls, ceilings and acoustic materials enable to host Eastern and Western repertoires. Special lighting effects coordinate activities with acoustics.

杭州大剧院首届迎新演出季
杭州歌舞团-迎新音乐会

杭州大剧院是杭州钱江新城第一座启动建造的建筑。

杭州市政府制定了一项城市总体规划，该规划包含一个国际会议中心、一个新政府大楼和一个大型购物中心。作为远离市中心的新城市商务中心，该规划计划提供土地给不同的潜在客户。

杭州大剧院被设计为一座城市的雕塑，是浙江省省会城市的主要文化设施之一。从空中俯瞰，剧院犹如月亮和珍珠。建筑由复杂的双曲面玻璃幕墙和钛合金屋顶组成。经典马蹄形的歌剧院大厅可容纳1600个座位，标准鞋盒式音乐厅可容纳600个座位，可变剧院可以容纳400个座位，实现8种观演模式变换，而屋顶24小时营业的餐厅可欣赏室外露天剧场和钱塘江的美景。现代前沿的舞台技术可兼顾东方和西方曲目，前置的建筑厅堂声学设计，及建造过程的精心管理，使杭州大剧院拥有极高的厅堂声学品质，测试结果几乎与设计数值一致。特殊的建筑泛光照明及效果使建筑与四周相协调，又独具风采。

WENZHOU GRAND THEATRE
温州大剧院

Cultural / 文化建筑

Wenzhou, Zhejiang, CHINA / 温州, 浙江, 中国

AWARDS / 获奖：1st Prize in International Design Competition / 国际设计竞赛一等奖
LOCATION / 地址：Fudong Rd., Lucheng District
CLIENT / 业主：Preparation Office of Wenzhou Grand Theatre
AREA / 面积：33 000 m^2
HEIGHT / 高度：46.06 m
DATE / 设计时间：2001
STATUS / 状态：Completed 2004 / 2004年建成
DESIGN ARCHITECT / 主创建筑师：Carlos Ott, PPA Architects
ARCHITECT OF RECORD / 合作方：Tongji University

Linking it to its maritime environment, the Wenzhou Theatre was inspired by a golden fish in a water pond. Strategically located, the theatre and main hall follow the main axis in the centre of the composition. Underneath it, the Concert Hall shares the main axis of the project. Offering a 1550-seat Opera House, a 650-seat Concert Hall, a 200-seat small theatre, restaurants and shops, the theatre also greets visitors with a generous outdoor plaza. The golden fish shape creates a unique form, revealing its tail marked by a stair over the main entrance, as a fresh splash of water. Its delicate image makes Wenzhou Grand Theatre a symbol of the city, easily recognizable around the world.

温州是一座沿海城市，温州大剧院的设计灵感来自一条跃起的锦鲤。歌剧院与入口大厅沿主轴依次分布，歌剧院正下方设置了音乐厅，同样沿主轴居中布置。歌剧院设有1550个座位，音乐厅有650个座位，小剧院有200个座位，此外还有一个宽敞的露天广场。整栋建筑外形似锦鲤，主入口处的大台阶正好成为锦鲤尾巴。整个建筑光彩夺目，呈现出一种独特的建筑造型。温州大剧院以其精美形象成为这座城市的景观，是具有世界级辨识度的建筑。

SEA & FOREST TOWER
海洋森林公寓

Residential / 住宅

Punta del Este, Maldonado, URUGUAY / 埃斯特角, 马尔多纳多, 乌拉圭

LOCATION / 地址：Parada 7, Mansa Beach
CLIENT / 业主：Sea Forest Group Development
AREA / 面积：12 800 m^2
HEIGHT / 高度：57 m, 23 storeys / 23层
UNITS / 户数：60
DATE / 设计时间：2001
STATUS / 状态：Completed 2004 / 2004年建成
DESIGN ARCHITECT / 主创建筑师：Carlos Ott
ARCHITECT OF RECORD / 合作方：Carlos Ott, Luis Rocca Architect

Mapocho
Av.
Los Alpes
VISITE UNIDADES DISPONIBLES
SEA FOREST GROUP
DEVELOPMENT
TRAUTMANN - GARCIA
Sea and Forest Tower

Overlooking the Playa Mansa of Punta del Este, with great views of the sea, the bay, the port, the Gorriti Island, the Playa Brava and the forest, the Sea Forest Tower emerges as an example of exquisite refinement. Elegance is injected in all 60 apartments; birds-eye-views are experienced from the suite bedrooms and terraced living rooms. Amenities offer tennis courts, spa, gym, playrooms, indoor and outdoor pools.Simplicity, vertical lines, repetition, a seaside curved facade and a magnificent top canopy give the tower an effective bold modern character.

该建筑俯瞰埃斯特角的曼萨海滩，在此可欣赏大海、海湾、港口、格里蒂岛、普拉亚布拉瓦河和森林的美景，海洋森林公寓是典雅精致住宅的典范。公寓中的60户住宅尽显优雅气质；从带独立浴室的卧室或带露台的起居室均可欣赏住宅外的美景。配套设施包括网球场、水疗中心、健身房、游戏室、室内和室外游泳池。简洁、直线、重复、海边弯曲的外墙和恢宏的挑檐使整个建筑呈现出大胆、现代的特征。

TIANJIN MARITIME COURTHOUSE
天津海事法院

Public / 公共建筑

Tianjin, CHINA / 天津, 中国

LOCATION / 地址：Second Ave. & Bei Hai Rd. East
AREA / 面积：12 300 m^2
HEIGHT / 高度：29.3 m
DATE / 设计时间：2001
STATUS / 状态：Completed 2004 / 2004年建成
DESIGN ARCHITECT / 主创建筑师：Carlos Ott, PPA Architects

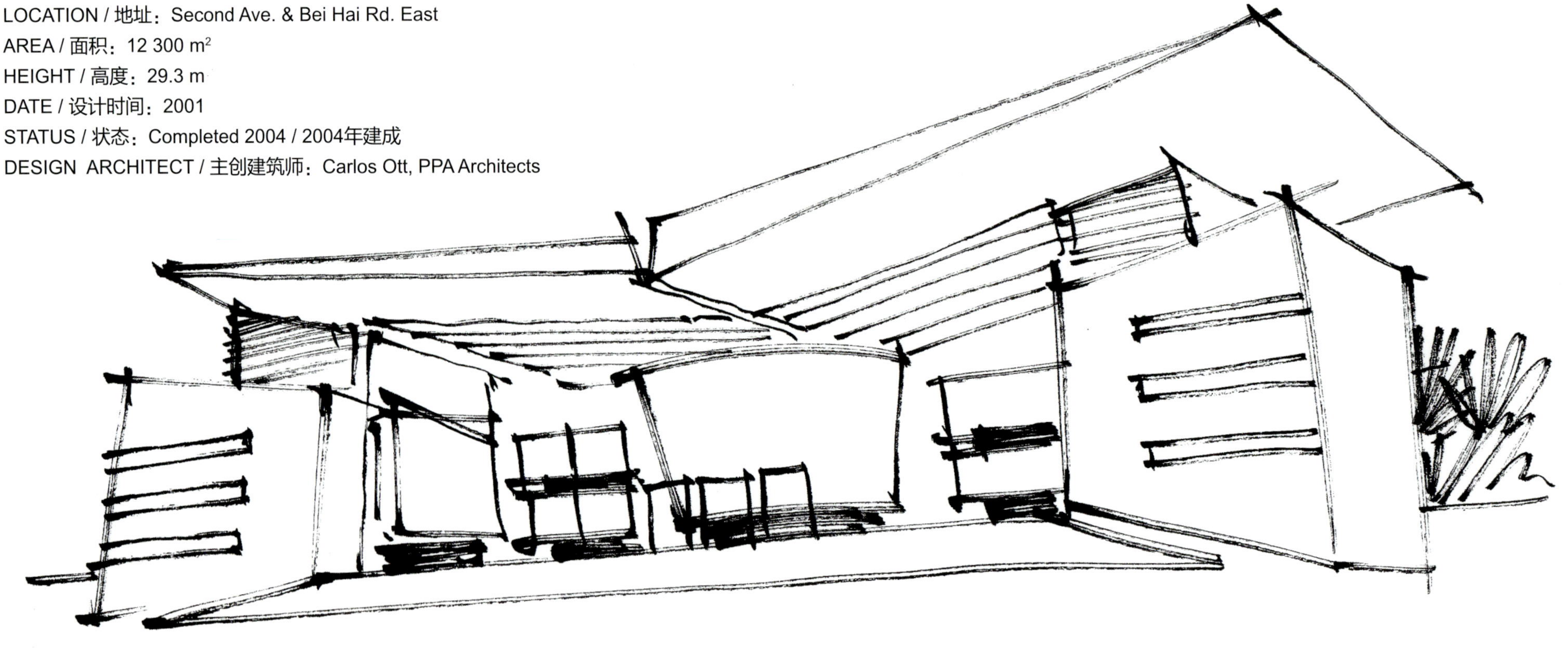

The Maritime Court displays a rigid layout and yet a flexible use. The multi-storey T-shaped luminous atrium is embraced by three rectangular solid constructions with slope walls. The effect of wall inclination makes the building even solemn and the glazed lobby enacts as a brilliant head crowned with an elegant but solid canopy.

天津海事法院建筑造型严谨庄重，但空间使用灵活。多层T形透亮中庭被三个带有倾斜壁的矩形实心结构所包裹。墙壁上收，使建筑物更显稳重，而玻璃大厅核心部位为明亮的中庭，屋顶造型优雅，犹如一本打开的法典。

FOUR POINTS SHERATON
福朋喜来登酒店

Hotel / 酒店

Dubai, UAE / 迪拜, 阿联酋

LOCATION / 地址：Sheikh Zayed Rd.

CLIENT / 业主：A. A. Almoosa Enterprises LLC

AREA / 面积：38 400 m^2

HEIGHT / 高度：166 m, 43 storeys / 43层

UNITS / 户数：194

DATE / 设计时间：2002

STATUS / 状态：Completed 2007 / 2007年建成

DESIGN ARCHITECT / 主创建筑师：Carlos Ott

ARCHITECT OF RECORD / 合作方：Arenco Architectural & Engineering Consultant

The Four Points Sheraton tower was designed preliminarily as a residential tower and re-defined as an important hotel and serviced apartments with 194 units afterwards.The 43rd-floor rooftop includes a swimming pool and spacious terrace with panoramic views of the Dubai skyline and the Jumeirah coastline. There is also a fitness centre and massage services. The hotel has 5 restaurants where guests can taste a wide variety of Italian, Moroccan and Indian cuisine. There is also a pub on the Mezzanine floor and a rooftop bar on the 43rd floor.

福朋喜来登酒店最初被设计为住宅楼，后被重新定义为一个重要的酒店和服务式公寓，共194户。43层的屋顶有游泳池和宽敞的露台，可以看到迪拜天际线和朱美拉海岸线的全景。酒店还提供健身中心和按摩服务。酒店有5家餐厅，住店客人可以品尝到各种各样的意大利、摩洛哥和印度美食，43楼还有一家屋顶酒吧。

DONGGUAN YULAN THEATRE
东莞玉兰剧院

Cultural / 文化建筑

Dongguan, Guangdong, CHINA / 东莞, 广东, 中国

AWARDS / 获奖：1st Prize in Design Competition / 设计竞赛一等奖

AREA / 面积：30 000 m^2

HEIGHT / 高度：39 m

DATE / 设计时间：2002

STATUS / 状态：Completed 2006 / 2006年建成

DESIGN ARCHITECT / 主创建筑师：Carlos Ott, PPA Architects

Like the gracious movements of a ballerina or the arm gestures of an orchestra conductor, the proposed image for the Dongguang Grand Theatre is of a curvilinear and dynamic dancing structure presenting the dramatic atmosphere of musical. The revolving conical volumes are clad in granite and glass, to differentiate the back stage and public zones, the large window of the foyer opening to the corner of the site provides a panoramic view of the grand mall.

东莞大剧院的形象是曲线的、动态的，宛如芭蕾舞演员旋转的舞姿，亦如交响乐指挥家挥动的手臂，似在用建筑形象表达建筑内的戏剧音乐活动。旋转的覆盖花岗岩的锥体嵌在玻璃体中，以区分后台和前场公共区域，门厅的大落地窗通向各角落，观众可在此欣赏大广场的全景。

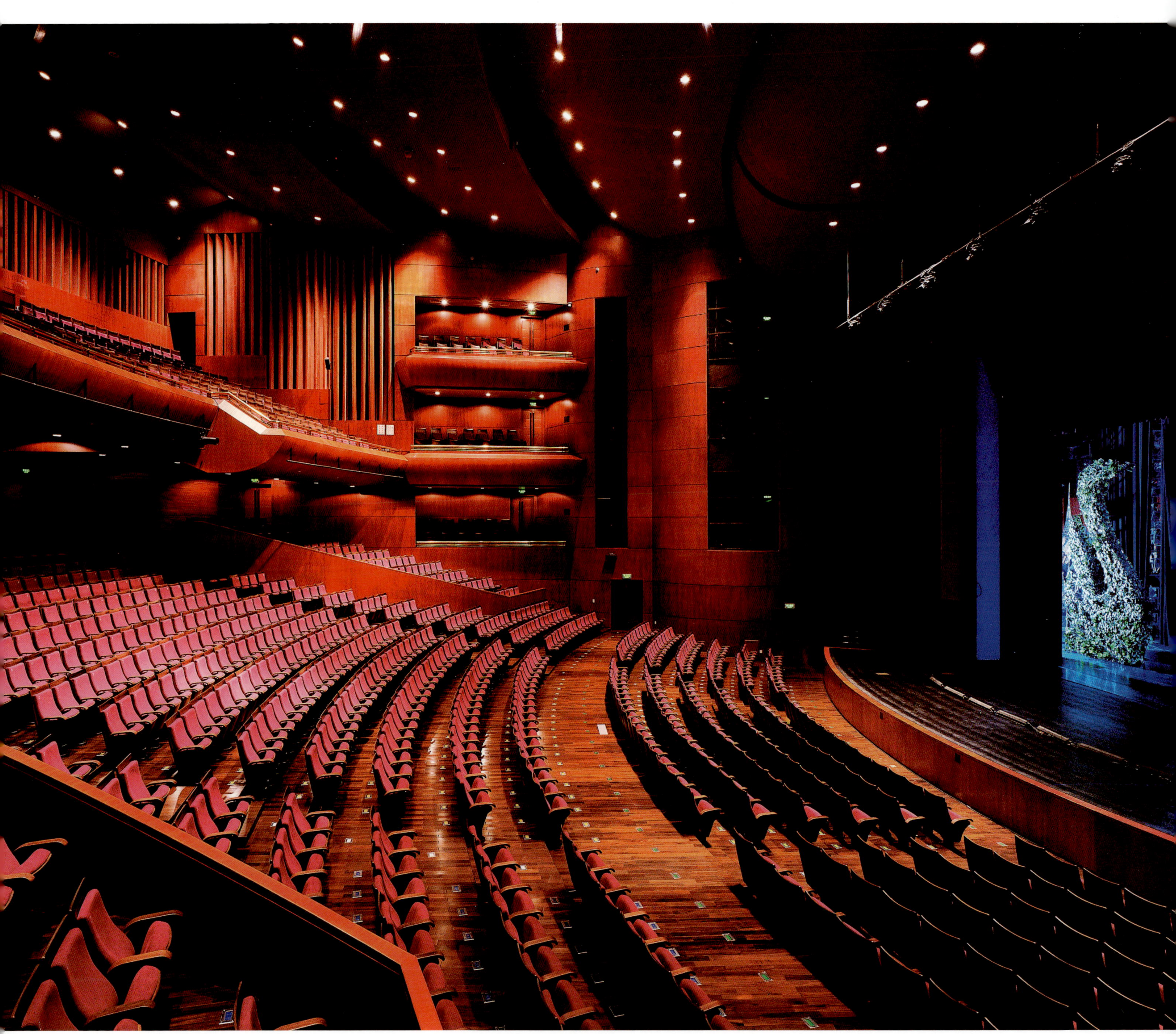

FALCON TOWER
Caesar Supermarket
NOW OPEN
Coca-Cola
SIGNATURE
GLASS
makes it cooler
FALCON TOWER

FALCON TOWER
猎鹰大厦

Residential / 住宅

Dubai, UAE / 迪拜, 阿联酋

LOCATION / 地址：Sh. Zayed Rd.

CLIENT / 业主：A. A. Almoosa Enterprises LLC

AREA / 面积：38 000 m^2

HEIGHT / 高度：163 m

DATE / 设计时间：2002

STATUS / 状态：Completed 2005 / 2005年建成

DESIGN ARCHITECT / 主创建筑师：Carlos Ott

ARCHITECT OF RECORD / 合作方：Arenco Architectural & Engineering Consultants

AAM TOWER
AAM大厦

Corporate / 企业总部
Dubai, UAE / 迪拜, 阿联酋

LOCATION / 地址：Sh. Zayed Rd.
CLIENT / 业主：A. A. Almoosa Enterprises LLC
AREA / 面积：101 615 m^2
HEIGHT / 高度：243.8 m
DATE / 设计时间：2002
STATUS / 状态：Completed 2008 / 2008年建成
DESIGN ARCHITECT / 主创建筑师：Carlos Ott
ARCHITECT OF RECORD / 合作方：Arenco Architectural & Engineering Consultants

An office tower with a dynamic form rises from a 35-metre-high parking podium. The architectural composition of this tower is based on the fusion of three triangular base prisms of different heights. The lower prism has a gold glass cladding. The intermediate height prism is embraced by silver glass. The highest prism is covered with an aluminium cladding.

一幢充满活力的办公大楼从一个35 m高的停车裙房上拔地而起。AAM大厦塔楼的建筑结构是三个不同高度的三角形棱柱：最矮的棱柱具有金色玻璃外墙，中间高度的棱柱是银色玻璃外墙，最高的棱柱是铝质外墙。

ARENCO
Call
800 8 2223
مخرج
EXIT 32
شارع النسيم
Al Naseem St
شارع المرسى
Al Marsa St
مخرج
EXIT

售票处

HENAN ART CENTRE
河南艺术中心

Cultural / 文化建筑

Zhengzhou, Henan, CHINA / 郑州, 河南, 中国

AWARDS / 获奖：1st. Prize in International Design Competition / 国际设计竞赛一等奖
LOCATION / 地址：Shangwu Inner Ring Rd., Jinshui District
CLIENT / 业主：Zhengdong New District Government Henan
AREA / 面积：63 000 m^2
HEIGHT / 高度：42 m
DATE / 设计时间：2003
STATUS / 状态：Completed 2008 / 2008年建成
DESIGN ARCHITECT / 主创建筑师：Carlos Ott, PPA Architects
ARCHITECT OF RECORD / 合作方：China Aviation Planning and Design Institute (Group) Co., Ltd.

The axes of these five ellipsoids of the Henan Art Centre merge into a same central point representing the significant geographic position of Zhengzhou. Two rising ax symmetric art walls reflect traditional spirit of Chinese ancient architecture and separate different zones as well. Multiple activities held in the gallery, the art museum and the theatre day and night, bring eternal vitality to this city.

河南艺术中心五个椭圆体的长轴汇集于一个中心，意寓着郑州市作为中原之中心的地理位置。两片翻卷上升的艺术墙为轴对称，赋予这组现代化建筑中国传统建筑的精髓，同时自然地将不同馆区划分开来，日间活动较多的美术馆、艺术馆与夜间活动较多的大剧院，为艺术中心带来昼夜不息的生气，又互不影响。

售票

Inspired by abstract artistic interpretations of ancient musical instruments, the centre fully embodies the concept of integrating ancient culture with modern arts. Variable heights and volumes of these halls allow them to be recognizable from outside which reinforces the asymmetrical concept within a symmetrical scheme. The patinated green copper resembles the color of jade stone considering to be a source of good luck.

五个椭圆体由河南出土的6500年前的古代乐器陶埙造型演变而来。艺术墙似黄河波涛翻卷的浪花，仿佛中国母亲河黄河，穿越并见证了中华上下五千年文明史；又如河南出土的2500年前的古代管乐器石排箫。中间晶莹剔透的装饰柱，是设计师根据河南出土的8700年前的中华第一笛贾湖骨笛而设计的。整个建筑群体均取自于古代乐器的抽象造型，使中原文化与现代建筑艺术有机地结合在了一起。

JADE BEACH
翡翠海滩公寓

Residential / 住宅

Sunny Isles Beach, Florida, USA / 阳光岛海滩, 佛罗里达, 美国

LOCATION / 地址：Collins Ave., Sunny Isles, Florida
CLIENT / 业主：Fortune International Group
AREA / 面积：73 000 m^2
HEIGHT / 高度：165.5 m
UNITS / 户数：248
DATE / 设计时间：2001
STATUS / 状态：Completed 2008 / 2008年建成
DESGIN ARCHITECT / 主创建筑师：Carlos Ott
ARCHITECT OF RECORD / 合作方：Revuelta Vega Leon P.A.

Becoming a landmark in the heart of Sunny Isles Beach, and enjoying bay, city and sea views, "Jade Beach" is the first tower of a pair; being a 51-storey tower with 248 spectacular residences. An impressive panoramic porte-cochere drop-off plaza overlooks both the city and the ocean. A dramatic entrance with impressive cascade fountains, lush stepping gardens, reflecting pools and Versailles-like staircases, leads to the 3-storey grand entrance lobby with breath-taking views of ocean and pool.

翡翠海滩公寓是两座塔楼中的第一座，享有海湾、城市和大海的美景，已成为阳光岛海滩中心的地标。这幢公寓高51层，拥有248户绝美住宅。令人印象深刻的全景门廊落客广场俯瞰着城市和海洋。建筑入口引人入胜，有令人印象深刻的喷泉、郁郁葱葱的阶梯花园和泛光的泳池。凡尔赛风格的楼梯，通向3层楼高的入户大厅，在大厅人们可直接欣赏到大海和游泳池的壮丽景色。

CALGARY COURTS CENTRE
卡尔加里联邦法院中心

Cultural, Offices / 文化建筑, 办公

Calgary, Alberta, CANADA / 卡尔加里, 阿尔伯塔, 加拿大

AWARDS / 获奖：Justice Facilities Review Awards, 2009, Citation winner, AIA´s Academy of Architecture for Justice / 2009年度司法建筑大奖（美国建筑师协会司法建筑学会颁发）

AIA Miami Chapter "Award of Merit" "Green Design", 2008 / 2008年度"优胜奖""环保设计奖"（美国建筑师协会迈阿密分会颁发）

LOCATION / 地址：7th Ave., 6th St, 5th St., 4th St., Calgary

CLIENT / 业主：Government of Alberta

AREA / 面积：110 000 m^2

HEIGHT / 高度：129 m

DATE / 设计时间：2003

STATUS / 状态：Completed 2007 / 2007年建成

DESGIN ARCHITECT / 主创建筑师：Carlos Ott, NORR

ARCHITECT OF RECORD / 合作方：Spillis Candela DMJM; NORR Architects, Engineers & Planners; Kasian Architecture Interior Design & Planning Inc.

The Project is a competition-winning design for a U$280 million development which will conform the largest consolidated Judicial Complex in North America. Two towers flank a soaring nearly atrium to create a landmark building for the Provincial Court, Court of Queens Bench and the Court of Appeal. This multilevel complex includes more than Courts, extensive Judicial Chambers, and rehabilitation of an existing heritage building and the creation of new urban park.

本项目的设计赢得了国际竞赛，项目耗资2.8亿美元，将与北美最大的综合司法综合体相接轨。卡尔加里法院中心是北美最大的综合司法机构之一，设有73个审判室和94个司法分庭。项目第一阶段由两幢楼组成，分别位于中央透明玻璃中庭的两侧。北侧高24层的塔楼设有皇后法院，南侧高20层的塔楼设有艾伯塔省法院。

20
21
19
18
18

2
CALGARY COURTS CENTRE

HANGZHOU INTERNATIONAL CONFERENCE CENTRE
杭州国际会议中心

Hotel, Convention Centre / 酒店, 会议中心

Hangzhou, Zhejiang, CHINA / 杭州, 浙江, 中国

LOCATION / 地址：Hangzhou CBD
CLIENT / 业主：Chief Engineer of Qianjiang CBD Area
AREA / 面积：114 400 m^2
HEIGHT / 高度：85 m, 18 storeys / 18层
DATE / 设计时间：2004
STATUS / 状态：Completed 2015 / 2015年建成
DESIGN ARCHITECT / 主创建筑师：Carlos Ott, PPA Architects
ARCHITECT OF RECORD / 合作方：China Aeronautical Project and Design Institute

Based on the principles of iconicity, functionality and innovation, the Hangzhou International Conference Centre adopts the spherical design concept of "sun", using the golden hue as the source of its main color while the curtain wall adopts a combination of modern composite panels and golden tinted glass. The entire building is magnificent looking: acting like a large golden sun, reflecting and accentuating the value of Hangzhou, solidifying the city´s reputation as a romantic capital.

杭州国际会议中心根据标志性、功能性、创新性的原则，采用球状"太阳"设计理念，以金色为主色调，幕墙采用现代材料复合板和金色玻璃相结合，整个建筑金碧辉煌，犹如一轮金色的太阳，体现杭州这座浪漫之都的价值取向。同时，杭州国际会议中心与半月形银色基调的杭州大剧院遥相呼应，一金一银，一日一月，一阴一阳，一缺一盈，天造地设一般。这组屹立在杭州城市最核心地带的城市建筑雕塑，不仅具有明确的象征意义，而且很好地暗示了杭州的过去和未来，在钱塘江畔相映生辉，使得"日月同辉"成为杭州的新地标。

ARTECH
阿特奇公寓

Residential / 住宅

Aventura, Florida, USA / 阿文图拉, 佛罗里达, 美国

LOCATION / 地址：2950 NE 188th St., Miami
CLIENT / 业主：Fortune International Group & Shefaor Development
AREA / 面积：58 700 m²
HEIGHT / 高度：27.15 m, 9 storeys / 9层
UNITS / 户数：232
DATE / 设计时间：2004
STATUS / 状态：Completed 2008 / 2008年建成
DESIGN ARCHITECT / 主创建筑师：Carlos Ott
ARCHITECT OF RECORD / 合作方：Charles Benson & Associates

Apartment complex Artech in Aventura is striking in its unusual shape, reminiscent of an ocean ship. Surrounded by tropical greenery, it offers residents an opportunity of enjoying the nature and the amenities of metropolis simultaneously.

位于阿文图拉的阿特奇公寓因其形状与众不同而引人注目，让人联想到一艘远洋轮船。它被热带植物所环绕，居民既能充分享受大自然的躁动又享有大都市的便利。公寓拥有多项便利设施，如水疗中心、健身中心、私人海滩及船坞，还可提供单层和跃层公寓。

JADE OCEAN
翡翠海洋公寓

Residential / 住宅

Sunny Isles Beach, Florida, USA / 阳光岛海滩, 佛罗里达, 美国

LOCATION / 地址：17121 Collins Ave., Sunny Isles, Florida
CLIENT / 业主：Fortune International Group
AREA / 面积：51 903 m^2
HEIGHT / 高度：167 m, 50 storeys / 50层
DATE / 设计时间：2004
STATUS / 状态：Completed 2009 / 2009年建成
DESIGN ARCHITECT / 主创建筑师：Carlos Ott
ARCHITECT OF RECORD / 合作方：Arc-Tech Associates Incorporated

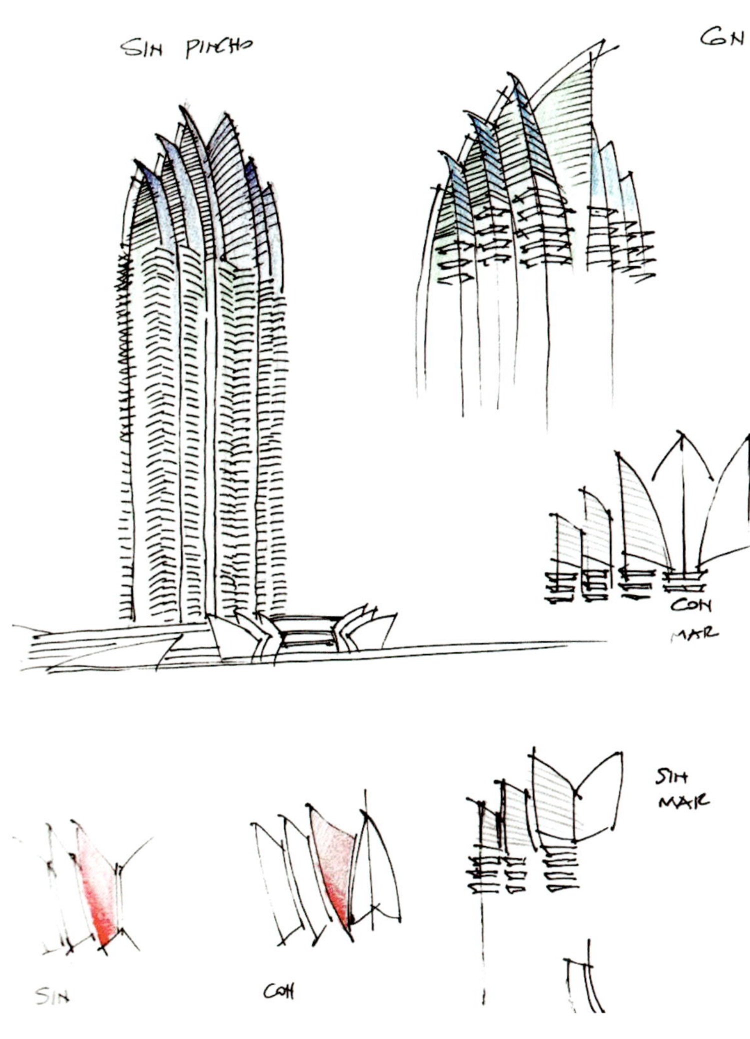

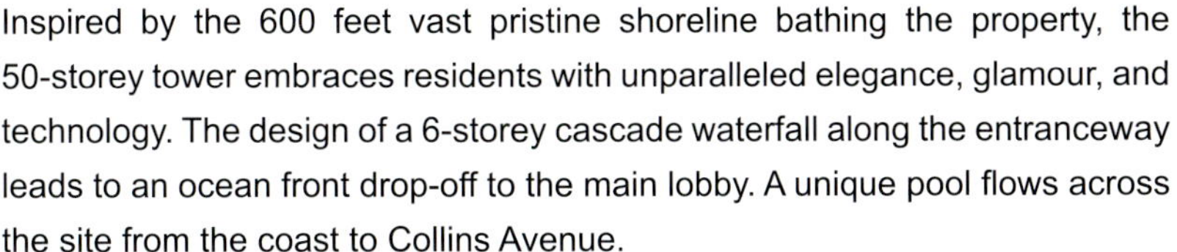

Inspired by the 600 feet vast pristine shoreline bathing the property, the 50-storey tower embraces residents with unparalleled elegance, glamour, and technology. The design of a 6-storey cascade waterfall along the entranceway leads to an ocean front drop-off to the main lobby. A unique pool flows across the site from the coast to Collins Avenue.

这座50层楼高的公寓紧临600英尺（182.88 m）的原始海岸线，以无与伦比的优雅、魅力和技术迎接住户。大堂入口设计有6层楼高的瀑布，通向海滨落客区。一个独特的游泳池从海岸延展过来，一直延伸到柯林斯大道。一系列豪华设施，包括落地玻璃墙、水疗中心、商务中心和最先进的技术，使其成为一座标志性建筑。

JADE BEACH & JADE OCEAN SALES CENTRE
翡翠海滩公寓和翡翠海洋公寓销售中心

Commercial / 商业

Sunny Isles Beach, Florida, USA / 阳光岛海滩, 佛罗里达, 美国

LOCATION / 地址：Collins Ave.，Sunny Isles, Florida
DATE / 设计时间：2005
STATUS / 状态：Completed 2005 - Demolished / 2005年建成，已拆除
DESIGN ARCHITECT / 主创建筑师：Carlos Ott

JADEOCEAN

DESIGN SUITES CALAFATE
卡拉法特设计套房酒店

Hotel / 酒店

El Calafate, Santa Cruz, ARGENTINA / 埃尔卡拉法特, 圣克鲁斯, 阿根廷

LOCATION / 地址：190 94th St., Playa Lago, Argentino
CLIENT / 业主：Design Suites
AREA / 面积：4600 m²
HEIGHT / 高度：13.6 m, 3 storeys / 3层
UNITS / 户数：60 rooms / 60间客房
DATE / 设计时间：2004
STATUS / 状态：Completed 2006 / 2006年建成
DESIGN ARCHITECT / 主创建筑师：Carlos Ott
ARCHITECT OF RECORD / 合作方：PWP Architects

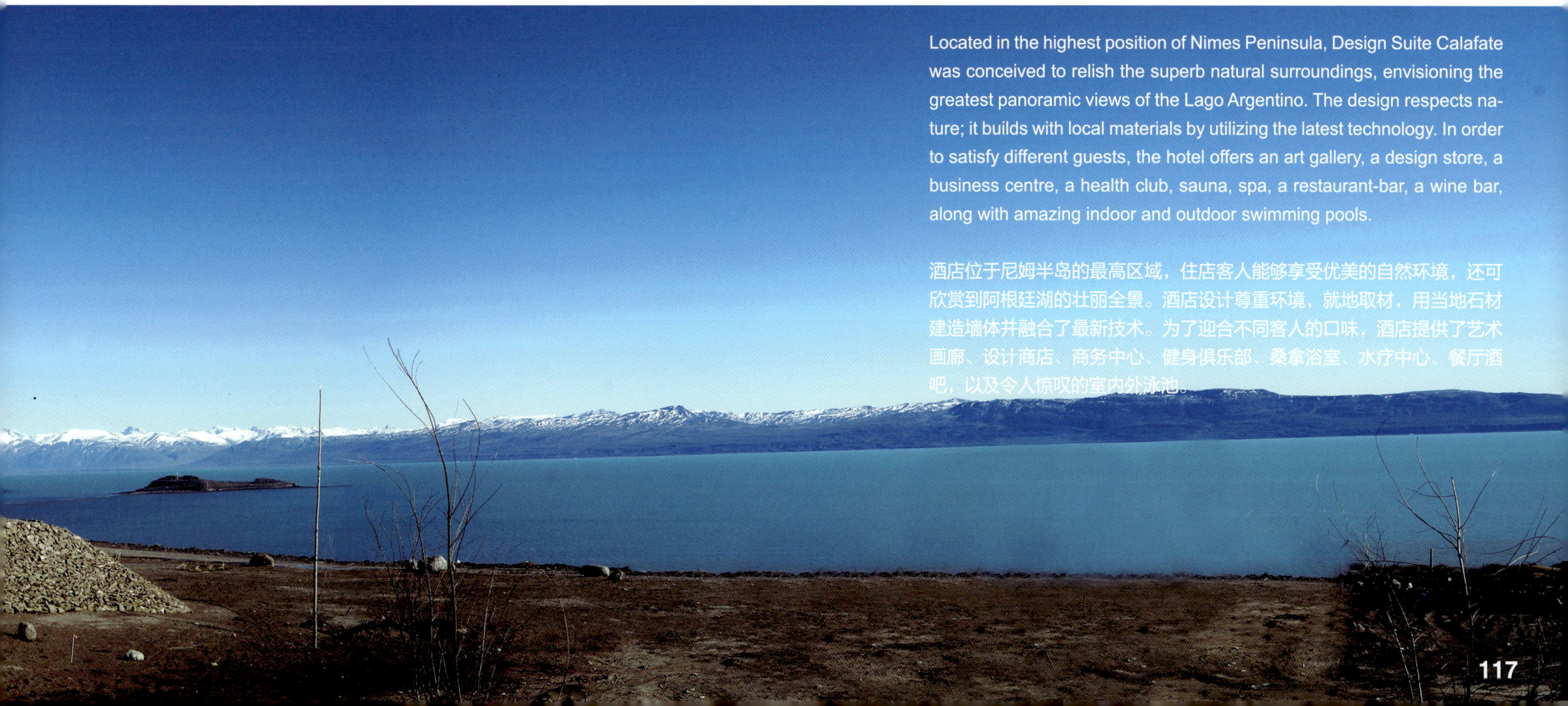

Located in the highest position of Nimes Peninsula, Design Suite Calafate was conceived to relish the superb natural surroundings, envisioning the greatest panoramic views of the Lago Argentino. The design respects nature; it builds with local materials by utilizing the latest technology. In order to satisfy different guests, the hotel offers an art gallery, a design store, a business centre, a health club, sauna, spa, a restaurant-bar, a wine bar, along with amazing indoor and outdoor swimming pools.

酒店位于尼姆半岛的最高区域，住店客人能够享受优美的自然环境，还可欣赏到阿根廷湖的壮丽全景。酒店设计尊重环境，就地取材，用当地石材建造墙体并融合了最新技术。为了迎合不同客人的口味，酒店提供了艺术画廊、设计商店、商务中心、健身俱乐部、桑拿浴室、水疗中心、餐厅酒吧，以及令人惊叹的室内外泳池。

TATA CONSULTANCY SERVICES SIRUSERI TECHNOLOGY PARK
塔塔集团科技园

Corporate / 企业总部

Chennai, Tamil Nadu, INDIA / 金奈, 泰米尔纳德, 印度

LOCATION / 地址：Mathematical Institute, Old Mahabalipuram Rd., Siruseri
CLIENT / 业主：Tata Consultancy Services (India)
AREA / 面积：400 000 m^2
DATE / 设计时间：2006
STATUS / 状态：Completed (Completion First Stage 2009, Second Stage 2011 and Third Stage 2016) / 一期2009年建成，二期2011年建成，三期2016年建成
DESIGN ARCHITECT / 主创建筑师：Carlos Ott ,Carlos Ponce de León Architects
ARCHITECT OF RECORD / 合作方: CRN Narayana Rao Architects & Engineers

The top software engineering TCS implements a spectacular technology park which covers around 28 000 hectares in Chennai located on the south-east coast of India, and is only 2 km away from the Indian Ocean. To architects, it is an incredible architectural opportunity for such a global leading company like TCS in this fantastic country. This project requires the architects to be experienced and sophisticated which makes the entire design process quite interesting and challenging.

作为印度软件业的旗舰，塔塔咨询服务公司在印度东南沿海的金奈建设了一个壮观的科技园，科技园占地面积280 km^2，距印度洋仅2 km。对于建筑师来说，在这个美丽的国度，为塔塔咨询服务公司这样的一流企业设计科技园，是非常难得的机会。这个项目需要建筑师具备纯熟的设计能力与丰富的设计经验，这些要求使得整个设计过程有趣且充满挑战。

UNION NATIONAL BANK

GAZELLE RESIDENCE
羚羊公寓

Residential, Commercial / 住宅，商业

Abu Dhabi, UAE / 阿布扎比，阿联酋

LOCATION / 地址：E48, Al Muntazah, Abu Dhabi

DATE / 设计时间：2005

STATUS / 状态：Completed 2009 / 2009年建成

DESIGN ARCHITECT / 主创建筑师：Carlos Ott

ARCADIA
阿卡迪亚公寓

Residential / 住宅

Montevideo, URUGUAY / 蒙得维的亚, 乌拉圭

LOCATION / 地址：Rambla República del Perú 1421
CLIENT / 业主：Sombra Fresca S.A.
AREA / 面积：10 370 m^2
HEIGHT / 高度：13 storeys / 13层
DATE / 设计时间：2005 – 2009
STATUS / 状态：Completed 2009 / 2009年建成
DESIGN ARCHITECT / 主创建筑师：Carlos Ott
ARCHITECT OF RECORD / 合作方：Raquel Rener Arch.

Located at the beachfront of Pocitos, one of the most luxurious areas in Montevideo, this sophisticated residence building integrates the newest intelligent technology. The distinctive facade is resolved through gold curtain glass, creating a curve along the entire height of the elevation. Balconies and railings in transparent glass allow a complete vision of the river with minimal cuts, making this building a landmark of Montevideo´s Rambla.

这座融合了最新智能技术的精致住宅楼位于蒙得维的亚最豪华地区之一的波西托斯海滨。整座大楼立面为独特的曲面，由金色玻璃幕墙构成。阳台栏杆由玻璃构成，用了最少的切割和遮挡，使楼外风景一览无余。这座建筑是蒙得维的亚兰布拉大街的地标。

VENTANAS AL PARAISO GRAND BAY
文塔纳斯帕莱索大海湾公寓

Residential / 住宅

San Francisco, Maldonado, URUGUAY / 圣弗朗西斯科, 马尔多纳多, 乌拉圭

CLIENT / 业主：Rafael Garfunkel-Daniel Binder
AREA / 面积：9800 m^2
HEIGHT / 高度：75 m, 4层
DATE / 设计时间：2005 – 2006
DESIGN ARCHITECT / 主创建筑师：Carlos Ott
ARCHITEC OF RECORD / 合作方：Carlos Ott, Edgar Baruzze Associated Architects

ONE SHENTON
珊顿一号公寓

Residential / 住宅

Singapore, SINGAPORE / 新加坡, 新加坡

AWARDS / 获奖：

FIABCI Prix d´Excellence Awards, 2014. World Gold Winner (Residential High Rise Category). — International Real Estate Federation / 2014年度世界卓越产业奖、世界金奖（高层住宅类）（世界不动产联盟颁发）

Construction Excellence Award, 2013 — Building and Construction Authority / 2013年度建筑卓越奖（新加坡建设局颁发）

Singapore Property Awards, 2012, Winner, Residential (High Rise Category) / 2012年度新加坡房地产评议会获奖（高层住宅类）

Green Mark Gold, 2008 — Building and Construction Authority / 2008年度绿色标志金奖（新加坡建设局颁发）

Best High-Rise Development, Singapore (Five-Star), 2008, CNBC Asia Pacific Property Awards / 2008年度英国CNBC国际房地产大奖亚太赛区之新加坡最佳高层开发五星大奖

LOCATION / 地址：Shenton Way & Commerce St.

CLIENT / 业主：City Developments Ltd.

AREA / 面积：72 000 m^2

HEIGHT / 高度：214 m; 45,50 storeys / 45,50层

UNITS / 户数：363

DATE / 设计时间：2005

STATUS / 状态：Completed 2011 / 2011年建成

DESIGN ARCHITECT / 主创建筑师：Carlos Ott

ARCHITECT OF RECORD / 合作方：Architects 61

Two residential towers located in Singapore´s downtown area, rise from a commercial podium. Gracious curvilinear gold and silver facades capture a variety of sunlights while appreciating the best views of the bay and the city. The podium, with stainless steel wire meshes cladding that camouflage the parking, continues the sinuosity of the towers. The spacious podium roof top accommodates the main swimming pool, the recreation and the landscaped areas. The bold asymmetrical towers are crowned by top roof decks with private pools and landscaped terraces.

位于新加坡市中心的两座住宅楼从商业裙房上升起。优美的金色曲线和银色外墙闪耀着每一缕阳光，同时还可欣赏海湾和城市的最佳景观。裙楼采用不锈钢丝网覆层，覆盖了停车场，延续了塔楼的蜿蜒曲折。宽敞的平台屋顶可容纳游泳池、休闲区和景观区。大胆的不对称塔楼由带有私人游泳池和景观露台的顶层结顶。

THE SOLITAIRE
纸牌公寓

Residential / 住宅

Singapore, SINGAPORE / 新加坡, 新加坡

AWARDS / 获奖：

Construction Excellence Award 2011 — Building and Construction Authority / 2011年度建筑卓越奖（新加坡建设局颁发）

Construction Productivity Award 2011 (Gold) — Building and Construction Authority / 2011年度建筑生产力金奖（新加坡建设局颁发）

Green Mark Platinum 2008 — Building and Construction Authority / 2008年度绿色标志白金奖（新加坡建设局颁发）

LOCATION / 地址：5A Balmoral Park, Tanglin

CLIENT / 业主：City Developments Ltd.

AREA / 面积：10 512 m²

HEIGHT / 高度：59 m, 12 storeys / 12层

UNITS / 户数：59

DATE / 设计时间：2005

STATUS / 状态：Completed 2011 / 2011年建成

DESIGN ARCHITECT / 主创建筑师：Carlos Ott

ARCHITECT OF RECORD / 合作方：Team Design Architects Pte Ltd.

Nestled in a quiet lush residential area, the Solitaire is a unique and iconic residential project, delineated in an attractive scheme for a challenging terrain and protected landscaping that offers a healthy living environment. Arranged in a terraced fashion down the slope and combined like piano keys, the buildings rise from the lush tropical landscape, providing breathtaking unobstructed views of the Botanic Gardens and Orchard Road for each and every apartment. External terraces, waterfalls and a large outdoor pool along with tennis court and pergolas, create a variety of outdoor experiences, in perpetual symbiosis with nature.

大楼坐落在一个安静的、郁郁葱葱的住宅区，是一个独特的标志性住宅项目，以极具挑战性的地形和受保护的景观设计引人注目，为住户提供健康的生活环境。这些建筑以梯田的方式沿着斜坡排列，像钢琴键一样组合在一起，从郁郁葱葱的热带景观中拔地而起，每一间公寓都可以欣赏植物园和乌节路美景。外部露台、瀑布和大型室外游泳池，以及网球场和凉棚架，创造了各种户外体验，与大自然交融。

VOLARI
沃拉里公寓

Residential / 住宅
Singapore, SINGAPORE / 新加坡, 新加坡

LOCATION / 地址：12 Balmoral Rd., Newton
DATE / 设计时间：2005
HEIGHT / 高度：53 m, 12 storeys / 12层
UNIT / 户数：85
STATUS / 状态：Completed 2014 / 2014年建成
DESIGN ARCHITECT / 主创建筑师：Carlos Ott
ARCHITECT OF RECORD / 合作方：Architects 61

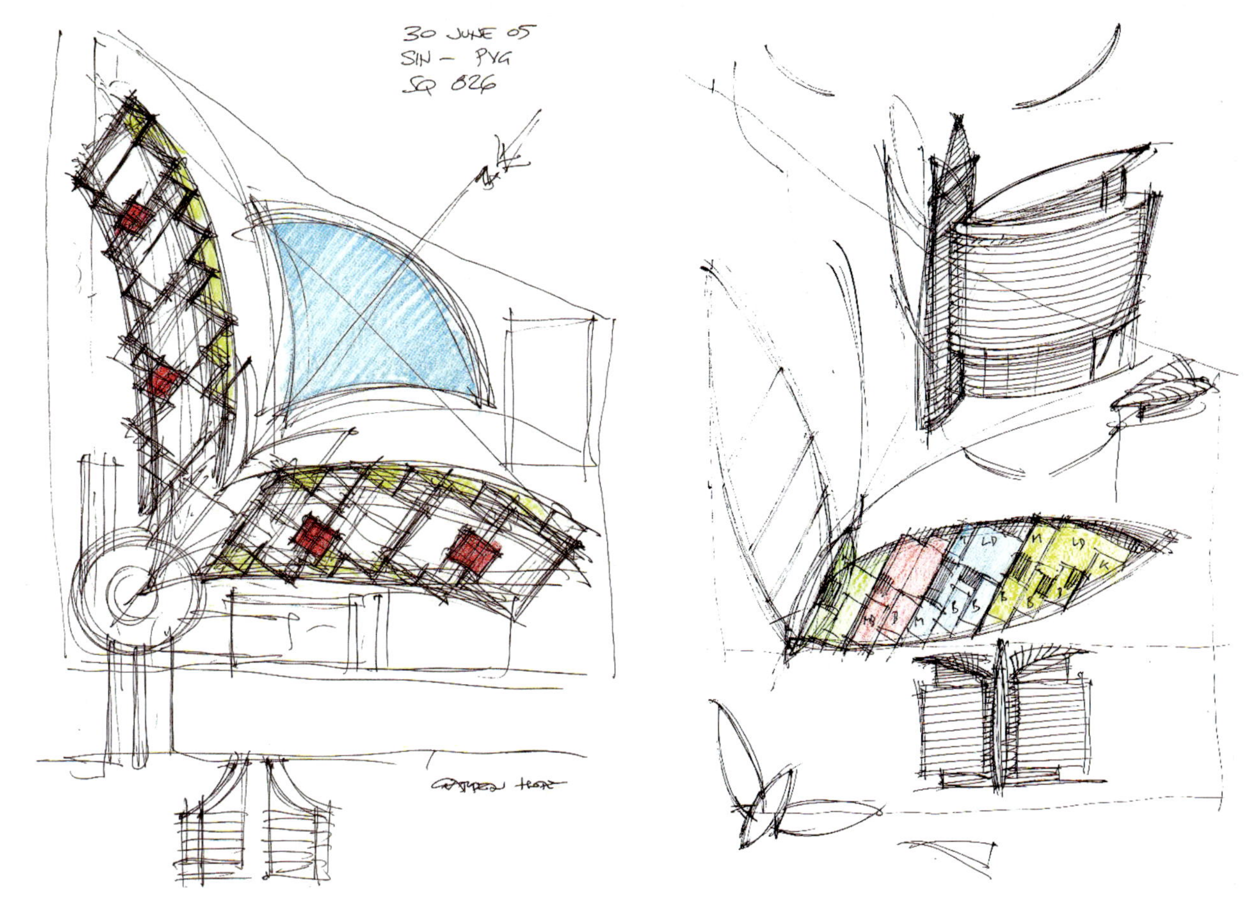
30 JUNE 05
SIN – PYG
SQ 826

This 85-unit residential building flies like a hang-glider over Nature, spreading its wings to embrace best orientation and views. Blue glass and shining aluminium presence coexist with stone concrete and greenery. Curved defiant design superimposes building structure and green surroundings, while introducing a sophisticated housing development.

这座拥有85个单元的住宅楼就像一架滑翔机飞越大自然，展开翅膀拥抱最佳方向和视野。蓝色玻璃和闪亮的铝材与石材混凝土、绿色植物相融合。曲线形的大胆设计将建筑结构和绿色环境叠加在一起，同时引入了复杂的住宅设计。

CLIVEDEN AT GRANGE
凯林豪庭

Residential / 住宅

Singapore, SINGAPORE / 新加坡, 新加坡

AWARDS / 获奖：

Construction Excellence Award 2013 — Building and Construction Authority / 2013年度建筑卓越奖（新加坡建设局颁发）

Design and Engineering Safety Excellence Award (2012)；Construction Productivity - Platinum (2012)；Universal Design - Silver (2012) — Building and Construction Authority / 2012年度优秀设计与工程安全奖、建筑生产力白金奖、通用设计银奖（新加坡建设局颁发）

Green Mark Platinum (2008)— Building and Construction Authority / 2008年度绿色标志白金奖（新加坡建设局颁发）

LOCATION / 地址：100 Grange Rd., River Valleys

CLIENT / 业主：City Developments Ltd

AREA / 面积：36 250 m^2

HEIGHT / 高度：24 storeys / 24层

UNITS / 户数：110

DATE / 设计时间：2005

STATUS / 状态：Completed 2011 / 2011年建成

DESIGN ARCHITECT / 主创建筑师：Carlos Ott

ARCHITECT OF RECORD / 合作方：ADDP Architects LLP

Cliveden at Grange reveals the secret of good turns. Sculptural blue glass towers of curvilinear geometry bloom from a lush garden like majestic ballerinas performing pirouettes in perfect synchrony. The mansions rise strikingly erect, emphasizing verticality, some arms stiff up in the air, others revolving gracefully. The bewildering arms of the dancers are formalized by spiral transparent glass balconies which gyrate accompanied by green planters. The powerful silver silhouettes of the upper braces provide refreshing shade, preserve intimacy with aluminum louvers and encircle the exquisite penthouse rooftops.

凯林豪庭塔楼的形态展示了旋转的优雅。几何曲线形状的雕塑般的蓝色玻璃塔从郁郁葱葱的花园中绽放，就像雄伟的芭蕾舞演员完美同步地表演旋转。塔楼形态垂直，一些“手臂”在空中坚挺，其他则优雅地旋转。舞者那令人眼花缭乱的手臂经由螺旋式透明玻璃阳台形式化，伴随着花架旋转。上部支架的银色轮廓营造了清晰的光影，与铝制百叶窗呼应，环绕在顶层公寓的屋顶上。

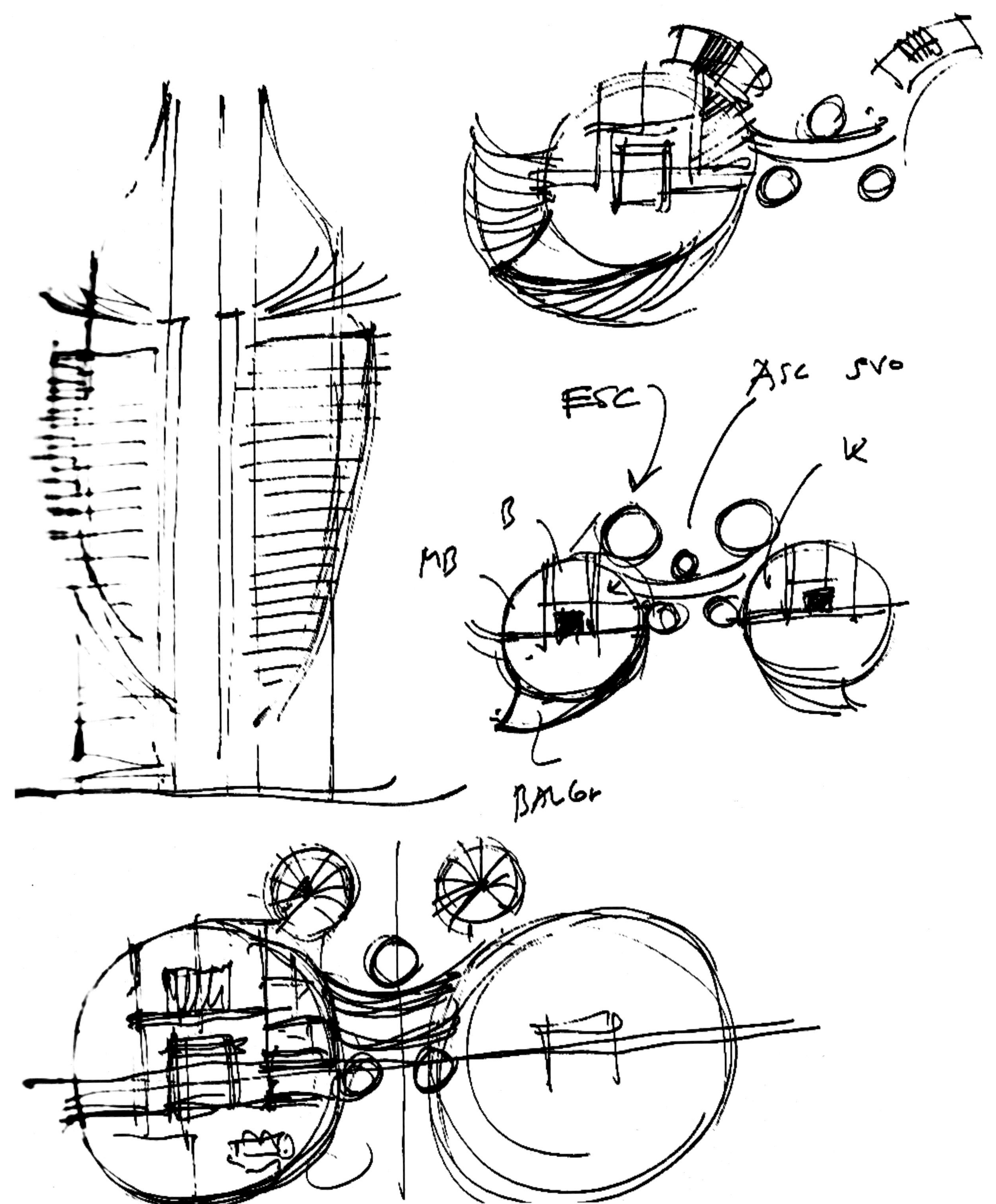
ESC
ASC SVO
K
B
MB
BALGr

MALVINAS ARGENTINAS INTERNATIONAL AIRPORT USHUAIA SECOND PHASE
乌斯怀亚国际机场二期

Airport / 机场

Ushuaia, Tierra del Fuego, ARGENTINA / 乌斯怀亚, 火地岛, 阿根廷

AWARDS / 获奖：1st Prize in International Design Competition / 国际设计竞赛一等奖
LOCATION / 地址：Ushuaia
CLIENT / 业主：London Supply S.A.
AREA / 面积：4000 m^2
HEIGHT / 高度：20 m
DATE / 设计时间：2006
STATUS / 状态：Completed 2009 / 2009年建成
DESIGN ARCHITECT / 主创建筑师：Carlos Ott
ARCHITECT OF RECORD / 合作方：Juan Carlos Sabaté

At the southern tip of South America, in close proximity to Antarctica, the International Airport of Ushuaia is considered the gateway to the remote "Tierra del Fuego" which often described as the "end of the world". Designed to symbolize the natural topography of the region, its sharp angles reflect the mountainous silhouette. The warm wood interior finishing contrasts with the rough natural stone walls giving a typical rustic ambiance in Tierra del Fuego.

乌斯怀亚国际机场位于南美洲的南端，紧邻南极洲，被认为是通往偏远的“火地岛”的门户，那里通常被称为“世界的尽头”。机场设计巧妙引用了该地区的自然地貌，其锐角反映了山区轮廓，温暖的木质内饰与粗糙的天然石墙形成鲜明对比，营造出阿根廷乌斯怀亚省的乡村氛围。

PLAYA VIK
维克宫精品酒店

Private Residence & Hotel / 私人住宅, 酒店
José Ignacio, Maldonado, URUGUAY / 何塞·伊格纳西奥, 马尔多纳多, 乌拉圭

LOCATION / 地址：Los Cisnes St.
CLIENT / 业主：Bermick S.A. - Carrie & Alexander Vik
AREA / 面积：7000 m^2
DATE / 设计时间：2006
STATUS / 状态：Completed 2010 / 2010年建成
DESIGN ARCHITECT / 主创建筑师：Carlos Ott
ARCHITECT OF RECORD / 合作方：Carlos Ott, Edgar Baruzze Associated Architects

A nonpareil spot on the peninsula of José Ignacio homes a magnificent beach retreat. Composed by a central structure surrounded by six guest houses. Conceived as a sculpture, a piece of art, the main house bears a particular curvilinear geometry. It appears to be based on the concept of the design of yachts, airplanes and sports cars, exposing its bold glass and titanium double curvature. Conveying a strong intention of respect towards the surrounding scales and proportions, the sculpture seems to be admiring the sea while reaching it with its grand longitudinal pool that cantilevers onto the beach.

何塞·伊格纳西奥位于一座半岛之上，其海滩是无与伦比的度假胜地。维克宫精品酒店由一个中央结构组成，周围有六间客房。主楼被视为雕塑、艺术品，具有特殊曲线的几何形状。它的设计理念来自游艇、飞机和跑车，大胆使用了玻璃和双曲钛金面板。该建筑尊重自然，亦与自然融合，在欣赏大海的同时，巨大的纵向游泳池还可延伸至海滩，直达大海。

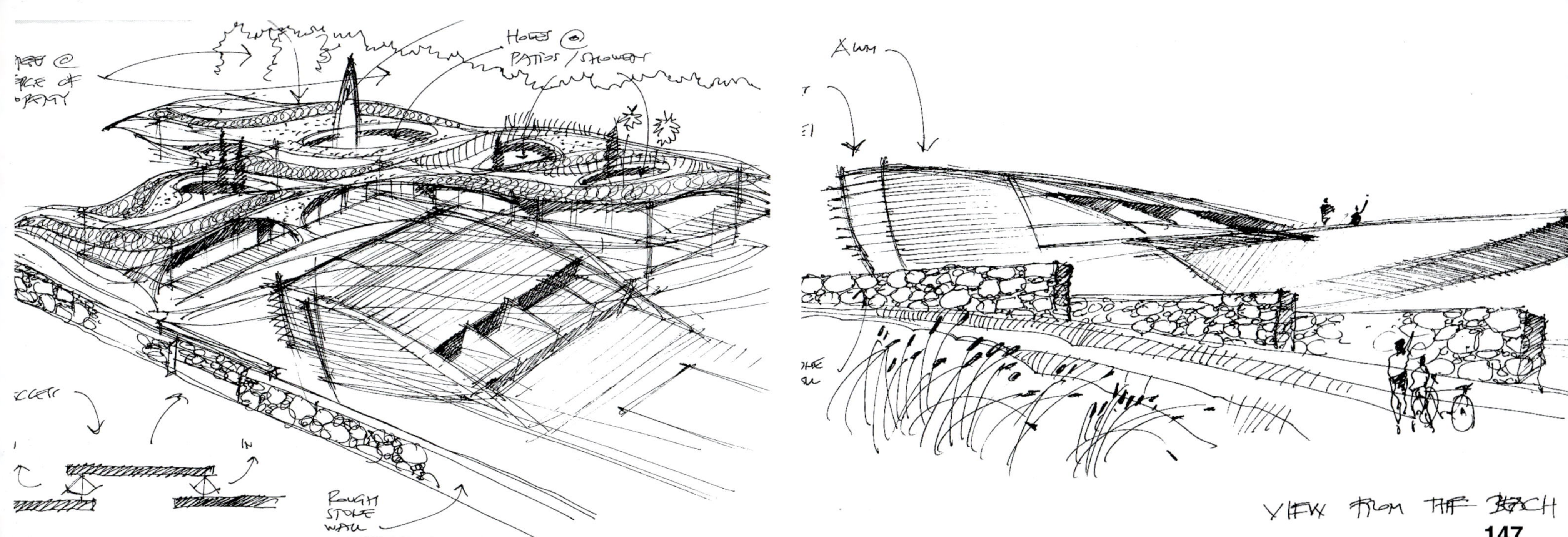
ROUGH STONE WALL
VIEW FROM THE BEACH

PARK LANE TOWER
柏宁酒店

Hotel, Office, Retail, Residential / 酒店, 办公, 零售, 住宅

Dubai, UAE / 迪拜, 阿联酋

LOCATION / 地址：Business Bay, Dubai
CLIENT / 业主：KM Properties
AREA / 面积：38 000 m^2
HEIGHT / 高度：151.8 m, 34 storeys / 34层
DATE / 设计时间：2006
STATUS / 状态：Completed 2018 / 2018年建成
DESIGN ARCHITECT / 主创建筑师：Carlos Ott
ARCHITECT OF RECORD / 合作方：Carlos Ott, Kling Consult

Aerodynamic in form, Park Lane is the bold image of a gravity defying structure. This business and residential tower, an extraordinary timeless landmark, is clad in aluminium and glass. Composed by a commercial podium, parking above and below grade, gym and spa, it offers highly efficient flexible office layouts on lower floors and 200 hotel rooms on upper floors. The day perception of this strong silhouette of refined architecture is strengthened at night by artistic lighting.

该建筑外形极具空气动力学特性，以及抵抗重力的大胆结构形象。这座商务和住宅大厦用铝和玻璃覆盖，是永恒的地标。它由商业裙房、地面及地下停车场、健身房和水疗中心组成，较低的楼层提供了高效灵活的办公空间，较高的楼层提供了200个酒店房间。夜间的艺术照明更加彰显了大楼的艺术造型。

YOO NORDELTA
YOO诺德塔公寓

Residential / 住宅

Tigre, Buenos Aires, ARGENTINA / 蒂格雷，布宜诺斯艾利斯，阿根廷

LOCATION / 地址：Nordelta District
CLIENT / 业主：Fideicomiso Condominios del Golf
AREA / 面积：40 000 m^2
HEIGHT / 高度：120 m
UNITS / 户数：110
DATE / 设计时间：2006
STATUS / 状态：Completed 2014 / 2014年建成
DESIGN ARCHITECT / 主创建筑师：Carlos Ott
ARCHITECT OF RECORD / 合作方：Carlos Ott, Christian Oppel Arquitectura

Positioned in a privileged location of Nordelta neighbourhood, with pathways that integrate the residences with the Golf Course designed by Jack Nicklaus, Yoo Nordelta blends architectural luxury and sophistication with the beauty of nature and the excitement of water sports that are characteristic of this magical spot.Aerodynamic aluminium roofs and curvilinear terraced glass facades, contrast with flanking stone walls.

Yoo诺德塔公寓坐落在诺德塔社区的优越位置，与杰克·尼克劳斯（Jack Nicklaus）设计的高尔夫球场仅一路相隔。该公寓将建筑的奢华和精致与自然之美和水上运动的刺激融为一体，成为其独具的特征。极具空气动力学特征的铝质屋顶和曲线形的露台玻璃幕墙，与石墙形成鲜明对比。

BUSINESS BAY OFFICES II
商务湾办公楼二期

Office / 办公

Business Bay, Dubai, UAE / 商务湾, 迪拜, 阿联酋

LOCATION / 地址：Sheikh Zayed Rd., Al Khail Rd.
CLIENT / 业主：KM Properties
AREA / 面积：18 200 m^2
HEIGHT / 高度：23 storeys / 23层
DATE / 设计时间：2006
STATUS / 状态：Completed 2015 / 2015年建成
DESIGN ARCHITECT / 主创建筑师：Carlos Ott
ARCHITECT OF RECORD / 合作方：Kling Consult (Hussein A. Shihab)

This building looks like a delicate blue diamond tulip tints the business Bay in downtown Dubai. The tower prospers from a calyx-like podium with several splits. The sepals arise dynamically enclosing the bewildering promenade along the covered walkway that surrounds the retail area. The structure, thus, is composed of a podium containing a retail floor, two parking floors and a technical floor, and nineteen flexible office storeys that indulges tenants with breathtaking views. Highly efficient, with multiple possibilities, the office layouts are able to serve either multiple or single tenants.The design is impressive, and this iconic blue diamond tulip blossoms in Dubai´s Business Bay day and night.

大楼形似蓝色钻石郁金香，为迪拜市中心的商务湾增色不少。塔楼从类似花萼的裙房上生长出来，萼片沿着包围零售区的走廊动态地包围着迷人的海滨长廊。该结构由包含零售、停车层和机房层的裙房，以及19层的办公楼组成，让租户可以欣赏到令人叹为观止的美景。办公室布局高效且具有多种可能性，可为多个或单个租户提供服务。建筑整体令人印象深刻，标志性的蓝色钻石郁金香在迪拜商务湾的白天绽放，在夜晚如烈火一样燃烧。

ALAMOS DE CARRASCO
卡拉斯科阿拉莫斯公寓

Residential / 住宅

Montevideo, URUGUAY / 蒙得维的亚, 乌拉圭

LOCATION / 地址：Bolivia Ave.
CLIENT / 业主：DEVCO
AREA / 面积：28 500 m^2
HEIGHT / 高度：5 storeys / 5层
UNITS / 户数：182
DATE / 设计时间：2007
STATUS / 状态：Completed 2014 / 2014年建成
DESIGN ARCHITECT / 主创建筑师：Carlos Ott
ARCHITECT OF RECORD / 合作方：Luis Rocca

Alamos de Carrasco, an attractive complex nested amidst a vast greenery that lies adjoining a park, is composed by a group of 5-storey residential buildings. These six curvilinear blocks and two towers contain 182 apartment units. There are 24 lofts, 48 one-bedroom apartments, 78 two-bedroom apartments and 32 three-bedroom apartments. Catering for a variety of ages, the complex also offers a pool and a Club House.

这是一组引人入胜的建筑群，坐落在毗邻公园的广阔绿地中，由一组5层高的住宅建筑组成。这6个曲线建筑和两座塔楼包含182间公寓，有24个跃层，48个一居室公寓，78个二居室公寓和32个三居室公寓。该住宅楼可满足各种年龄段住户的需求，还设有一个游泳池和一间俱乐部会所。

LAGUNA ESCONDIDA BEACH & LAGOON
拉古纳埃斯康迪达酒店

Hotel & Residential Villas / 酒店，度假村

José Ignacio, Maldonado, URUGUAY / 何塞·伊格纳西奥，马尔多纳多，乌拉圭

LOCATION / 地址：Ruta 10, Arenas de José Ignacio
CLIENT / 业主：Related Group
AREA / 面积：28 500m^2
UNTIS / 户数：227
DATE / 设计时间：2007
STATUS / 状态：Completed 2014 / 2014年建成
DESIGN ARCHITECT / 主创建筑师：Carlos Ott

Laguna Escondida Beach & Lagoon is an exclusive gated summer resort in José Ignacio, the top seaside location on the Atlantic Ocean in Punta del Este, Uruguay. The masterplan includes beach service, a boutique hotel in the bay, ocean views surrounded by the forest, with countryside peace and lagoon serenity.

酒店位于何塞·伊格纳西奥绝佳的避暑胜地，在乌拉圭埃斯特角城大西洋海边的最高位置。总体规划包括海滩服务设施、海湾中的精品酒店，酒店被森林包围并享有海景，具有乡村的宁静和泻湖的神秘。

NATIONAL BANK OF ABU DHABI
阿布扎比国家银行

Commercial / 商业

Abu Dhabi, UAE / 阿布扎比, 阿联酋

LOCATION/地址：Sheikh Khalif St., Abu Dhabi
CLIENT / 业主：National Bank of Abu Dhabi
AREA / 面积：16 139 m^2
HEIGHT / 高度：91 m, 12 storeys / 12层
DATE / 设计时间：2007
STATUS / 状态：Completed 2012 / 2012年建成
DESIGN ARCHITECT / 主创建筑师：Carlos Ott
ARCHITECT OF RECORD / 合作方：APG Architecture & Planning Group

This headquarters for the new branch of the National Bank of Abu Dhabi will reinforce the emblematic image set by the existing NBAD, with the double tower structure in pyramidal cross-section with inversely reflected triangular peaks. Clad in blue reflective glass, the pyramid over the entrance, crowns both, the clear glass lobby and the walkway that is defined by bold black granite columns. The second pyramid is clad in grey reflective glass, giving a tint of solemnity for the banking centre. Analogous to its predecessor, this new branch will definitely augment the iconic shape that is already implanted in Abu Dhabi in relation to this prestigious Bank.

阿布扎比国家银行新分行的最新总部具有双塔结构的金字塔形横截面和倒映的三角形山峰，加强了现有阿布扎比国家银行设定的标志性形象。入口上方的金字塔覆盖着蓝色反光玻璃，罩在透明玻璃大堂和黑色花岗岩柱子的走道上。第二个金字塔覆盖着灰色反光玻璃，为银行中心增添了一丝庄严。与其他分行类似，这家新分行增强了已深入人心的阿布扎比国家银行的标志性形状，强调着这家享有盛誉的银行的形象。

LE BLEU
蓝色公寓

Residential / 住宅

Punta del Este, Maldonado, URUGUAY / 埃斯特角，马尔多纳多，乌拉圭

LOCATION / 地址：Port Rambla Artigas
CLIENT / 业主：Vedopark S.A.
AREA / 面积：2600 m^2
HEIGHT / 高度：16 m
DATE / 设计时间：2008
STATUS / 状态：Completed 2011 / 2011年建成
DESIGN ARCHITECT / 主创建筑师：Carlos Ott
ARCHITECT OF RECORD / 合作方：Vicente Aramendia

Le Bleu is a residential building developed into five exclusive units, one per level, with glazed private pools on their terraces viewing sunsets over Punta del Este harbour and Gorriti Island. The building entrance on the rear 20th Street is composed by a fully glazed multiple height luminous main lobby with a panoramic lift, leads to the private lobby along a light transparent bridge hung among Zen courtyards. The parking entrance is along the Rambla, while the panoramic lift leads to the meeting room and the fitness centre.The roof top deck furnished with a barbecue, a solarium and a counter current lap pool.

该公寓是一栋住宅建筑，分为5个豪华单元，每层一个，在露台上设有玻璃私人泳池，可欣赏埃斯特角港和格里蒂岛的日落。建筑物入口位于第20街后方，由全玻璃、高透光的主大堂和全景电梯组成，沿着吊在禅宗庭院之间的透光桥通向私人大堂。停车场入口在兰布拉大街，而全景电梯则通往会议室和健身中心。屋顶露台配有烧烤设施、日光浴室和小型游泳池。

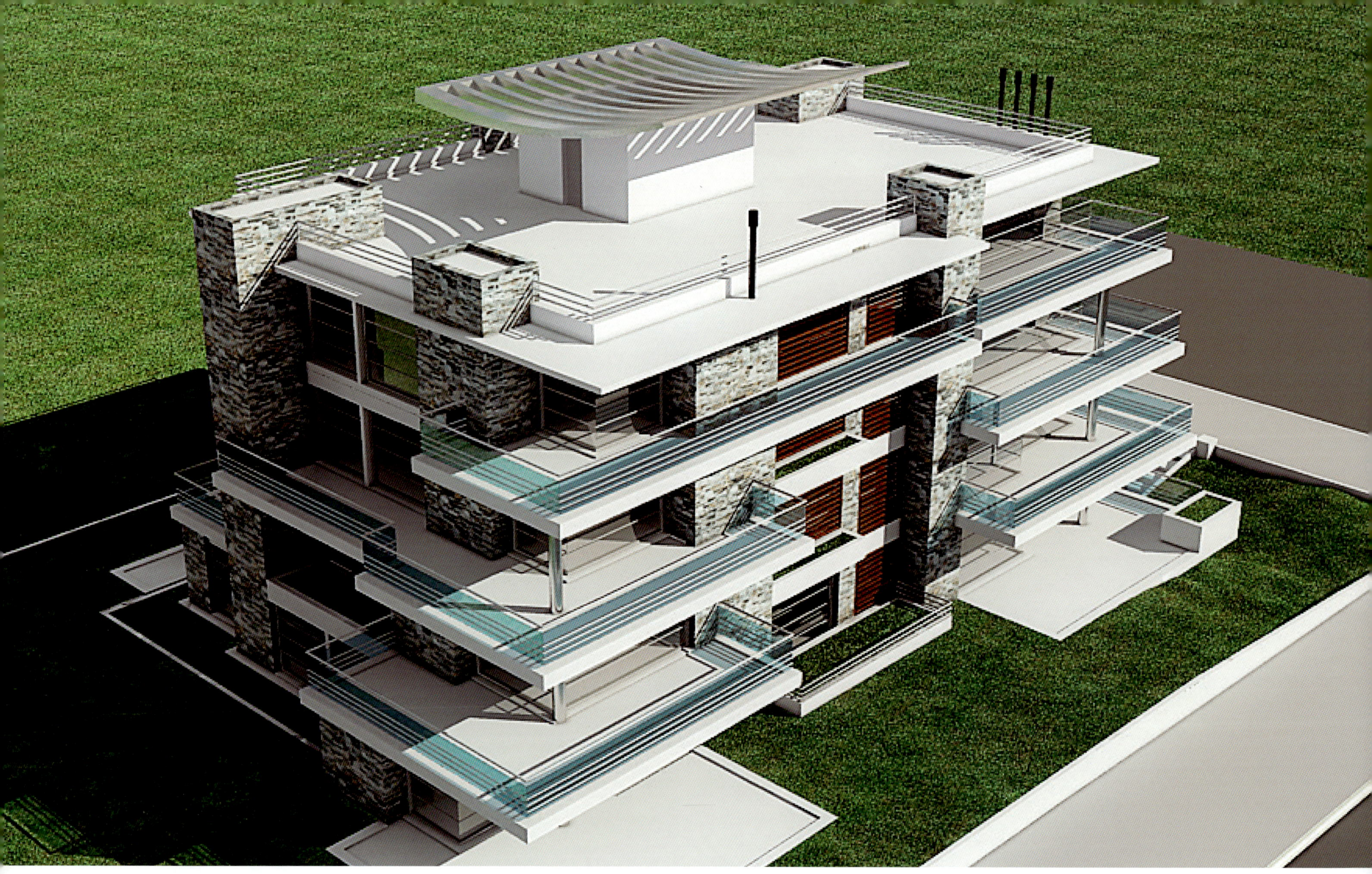

LAWN RESIDENCES
草坪公寓

Residential / 住宅

Montevideo, URUGUAY / 蒙得维的亚, 乌拉圭

LOCATION / 地址：Leonel Aguirre 1750, Carrasco, Montevideo
CLIENT / 业主：Fideicomiso Harwood
AREA / 面积：1817 m^2
HEIGHT / 高度：12 m
UNITS / 户数：5
DATE / 设计时间：2008
STATUS / 状态：Completed 2011 / 2011年建成
DESIGN ARCHITECT / 主创建筑师：Carlos Ott
ARCHITECT OF RECORD / 合作方：Edgar Baruzze Associated Architects

Located in an exclusive residential neighbourhood of Montevideo, close to the Carrasco Lawn Tennis Club, the luxurious 4-storey Lawn Residences contain five apartments, developed to enhance park views from its open terraces. Flagstone, aluminium and glass define this construction, while merging with the environment.

该建筑位于蒙得维的亚一个高级住宅区，靠近卡拉斯科草坪网球俱乐部，共包含5套公寓，设计意图是通过开放式露台增进与外部公园美景的互动。石板、铝板和玻璃为建筑的主要材质，同时又与环境融为一体。

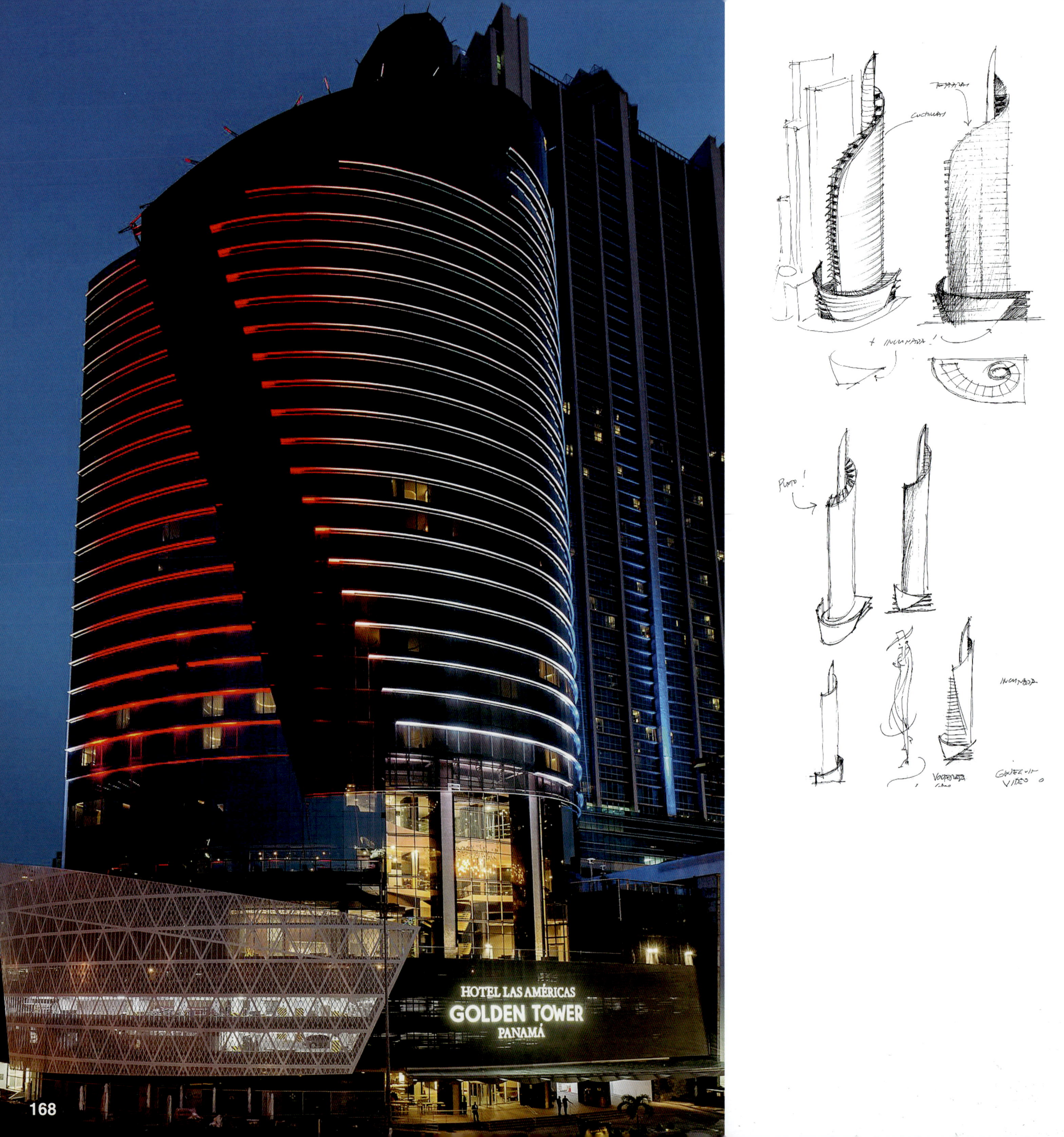
HOTEL LAS AMÉRICAS
GOLDEN TOWER
PANAMÁ

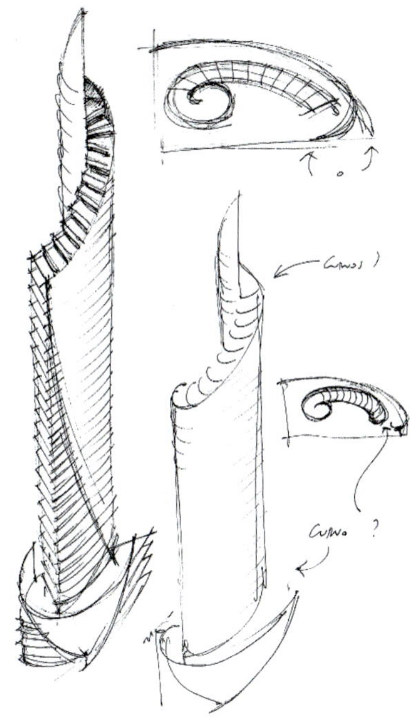

LAS AMERICAS GOLDEN TOWER PANAMA HOTEL
巴拿马美洲金塔酒店

Hotel / 酒店

Panama, PANAMA / 巴拿马, 巴拿马

LOCATION / 地址：Balboa Ave. & 53 St.
CLIENT / 业主：Olympix Trading S.A.
AREA / 面积：36 000 m^2
HEIGHT / 高度：153.5 m, 38 storeys / 38层
UNITS / 户数：267 rooms, 18 suites / 267间客房，18间套房
DATE / 设计时间：2008
STATUS / 状态：Completed 2016 / 2016年建成
DESIGN ARCHITECT / 主创建筑师：Carlos Ott
ARCHITECT OF RECORD / 合作方：Eduardo Delgado

A magnificent golden helicoid curls to the sky to host this 285-key hotel. Conceived as a five-star hotel and business centre, the project offers all the facilities and comfort to guests. The entrance hall connects directly to the check-in lobby. The podium offers 47 parking stalls on three levels. Level +13.4 m is destined for hotel customers, offering financial areas, a conference centre, a ballroom, and a boardroom along with administration and executive offices.

这家拥有285间客房的酒店如同宏伟的金色螺旋，卷曲着伸向天空。建筑包含五星级酒店和商务中心，为客人提供设施和舒适办公环境。一楼入口大厅直通登记大厅。裙楼提供47个停车位。+13.4 m层专供酒店客户使用，提供金融区、会议中心、宴会厅和会议室以及行政和执行办公室等。

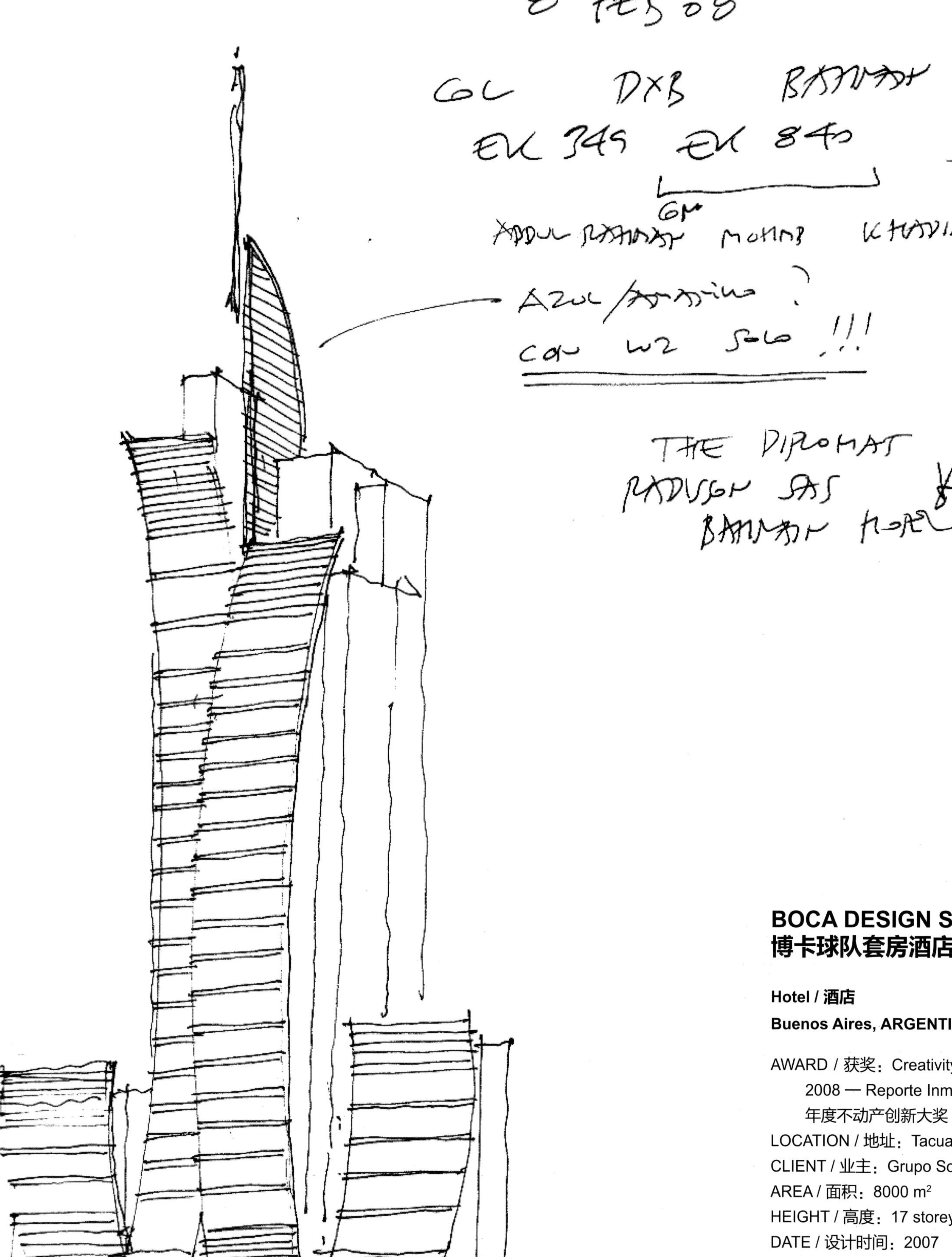

BOCA DESIGN SUITES HOTEL
博卡球队套房酒店

Hotel / 酒店

Buenos Aires, ARGENTINA / 布宜诺斯艾利斯, 阿根廷

AWARD / 获奖：Creativity and Innovation in Real Estate Award 2008 — Reporte Inmobiliario Economia & Real State / 2008 年度不动产创新大奖（阿根廷房地产报告网站颁发）
LOCATION / 地址：Tacuari 243, Buenos Aires
CLIENT / 业主：Grupo Solanas
AREA / 面积：8000 m^2
HEIGHT / 高度：17 storeys / 17层
DATE / 设计时间：2007
STATUS / 状态：Completed 2012 / 2012年建成
DESIGN ARCHITECT / 主创建筑师：Carlos Ott
ARCHITECT OF RECORD / 合作方：PWP Architects

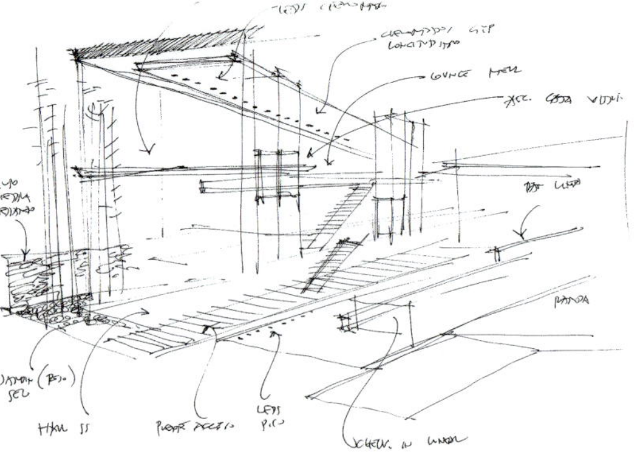

Located in downtown Buenos Aires, this 85-suite, 17-storey Design Hotel is conceived to become an icon of this perpetually avant-garde city. All four facades are developed with a unique design generating an additional visual interest. The main entrance facade is composed of four glazed curved surfaces establishing a dialogue in between. The lower lateral arched planes, amalgamate with the surrounding buildings. The main lobby is defined as a permanent visual reference, which is vertically connected by a glass box containing elevators. Ground floor includes the entrance, lobby and dining. First basement offers the multi-purpose room. First level accommodates the spa. These three levels interconnect by a triple height spacious hall that establishes a close connection with the street and its flow.The design of Boca Juniors is reflected in the longitudinal disposition of long and narrow panels of flooring, ceiling and cladding. The linear sequence becomes the leitmotif of the composition. The main space thus is conceived as a journey that begins when you enter the hotel, integrating visual, technical and emotional aspects.

这座拥有85间套房、17层高的酒店位于布宜诺斯艾利斯市中心，旨在成为这座城市的标志。所有四个立面均采用独特的设计，产生不同的视觉效果。主入口立面由四个玻璃曲面组成，相互呼应。较低的横向拱形平面在高度上与周围建筑物融为一体。主大厅由一个包含电梯的玻璃盒子垂直连接，该设计最终成为一个经典的设计参考。一楼包括入口、大堂和餐厅。地下一层提供多功能室。第一层有水疗中心。这三个楼层通过一个三层高的宽敞大厅相互连接，与街道建立了密切的联系。关于博卡青年竞技俱乐部的设计体现在酒店内饰上，如地板、天花板和覆层的长而窄的面板设计。线性序列成为构图的主旋律，人们从进入酒店开始就犹如开启一段旅程，整个设计融合了视觉、技术和情感，令人惊叹。

KAMALA TOWER
卡马拉大厦

Commercial / 商业

Al Khalidiya, Abu Dhabi, UAE / 哈利迪亚，阿布扎比, 阿联酋

LOCATION / 地址：Sheikh Zayed The First St.
CLIENT / 业主：Sheikh Tahnoon bin Zayed al Nahyan
AREA / 面积：17 000 m^2
HEIGHT / 高度：90 m, 23 storeys / 23层
DATE / 设计时间：2008
STATUS / 状态：Completed 2017 / 2017年建成
DESIGN ARCHITECT / 主创建筑师：Carlos Ott
ARCHITECT OF RECORD / 合作方：Herberger Engineering

Discreet elegance is embodied in this curvilinear clear blue tower, shielded by splendid golden glass planes and crowned with bold black granite. Commercial areas are defined in ground floor and mezzanine, while office areas are hosted in 19 storeys of the tower. A grandiose aluminium canopy invites visitors to the main lobby, while maintaining a dialogue with the top canopy that covers the roof deck terrace that belongs to the peak office.

这座蓝色大楼的形体曲线克制而优雅，立面为华丽的金色玻璃，顶部为黑色花岗岩材质。建筑底层和夹层为商业区，上部19层为办公区域。建筑主入口上方设置巨大的悬挑金属雨篷，与塔楼顶部的悬挑构架遥相呼应。

THE NEW EMI STATE TOWER
新EMI集团大厦

Residential & Offices, Commercial / 住宅, 办公, 商业
Abu Dhabi, UAE / 阿布扎比, 阿联酋

LOCATION / 地址：Hamdan Bin Mohammed St. & Airport Rd., Central Souk, Zone 1E3-01
CLIENT / 业主：SH. Tahnoon Bin Saeed Bin Shakhbott Al Nahyan
AREA / 面积：42 000 m^2
HEIGHT / 高度：153.5 m, 32 storeys / 32层
DATE / 设计时间：2008
STATUS / 状态：Completed 2016 / 2016年建成
DESIGN ARCHITECT / 主创建筑师：Carlos Ott
ARCHITECT OF RECORD / 合作方：Crang & Boake Inc., World Planners Inc.

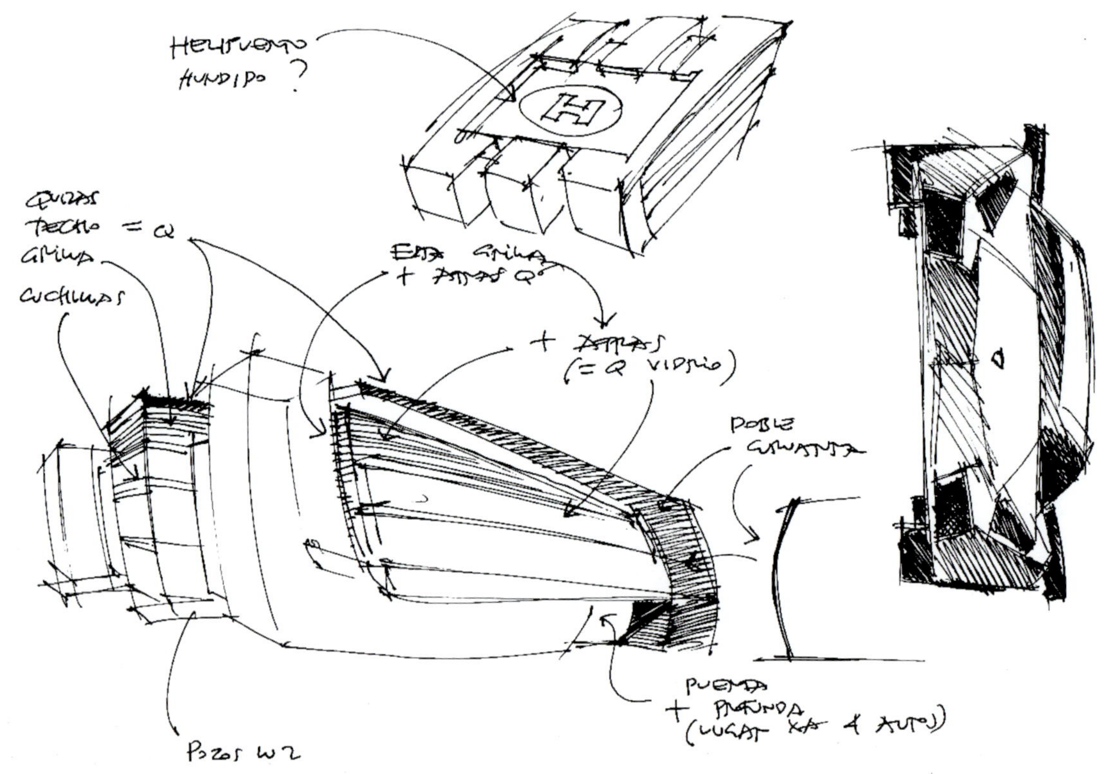

LAMBORGHINI DUBAI
兰博基尼迪拜展厅

Showroom, Commercial / 品牌销售展厅，商业

Dubai, UAE / 阿布扎比, 阿联酋

LOCATION / 地址：Sheikh Zayed RD Exit 41, Umm Al Sheif
CLIENT / 业主：Al Jaziri Group
AREA / 面积：13 600 m^2
HEIGHT / 高度：20.2 m, 3 storeys / 3层
DATE / 设计时间：2008
STATUS / 状态：Completed 2017 / 2017年建成
DESIGN ARCHITECT / 主创建筑师：Carlos Ott
ARCHITECT OF RECORD / 合作方：Design Center Architects & Engineers Consultants

Lamborghini Dubai, the world's largest brand showroom and dedicated service centre, is located on the Emirate's main thoroughfare Sheikh Zayed Road. The aerodynamic design based on curvilinear lines, clear and grey glass, black steel cladding, dark hexagonal honeycomb metal grids, metallic louvers and brilliant white fins; expresses the audacious language of Lamborghini. The iconic 3-storey showroom and office construction emphasizes the spirit of the automobile line, designed to cater for select, specialized audiences.

位于阿联酋主要通道谢赫·扎耶德路上的兰博基尼迪拜是世界上最大的品牌展厅和专业服务中心，具空气动力学特征的曲线形设计，透明的灰色玻璃，黑色的钢质外覆层，深色的六角形蜂窝状金属网格，金属百叶窗和明亮的白色散热片都传达着兰博基尼的大胆。标志性的三层展示厅和办公厅彰显了品牌精神，专为有品位的客户服务。

CELEBRA BUILDING
塞勒贝拉大楼

Corporate / 企业总部

Montevideo, URUGUAY / 蒙得维的亚, 乌拉圭

AWARDS/获奖：International Property Award in association with Bloomberg television: 1st Architecture Award (office) category for Uruguay, 1st Architecture Award (office) category for America, 1st International Architecture Award (office) / 国际地产奖（彭博电视台主办）：乌拉圭建筑大奖第一名（办公类）、美洲地区建筑大奖第一名（办公类）、国际建筑大奖第一名（办公类）

LEED 2009 Core and Shell GOLD Certified 2015 USGBC / 2015年度美国绿色建筑委员会LEED标准（2009版）认证：主体结构及外围护系统金级

LOCATION / 地址：Km 17.500, Ruta 8, Zonamerica Free Zone

CLIENT / 业主：Zonamerica S.A.

AREA / 面积：10 324 m^2

HEIGHT / 高度：38.6 m, 8 storeys / 8层

DATE / 设计时间：2009

STATUS / 状态：Completed 2014 / 2014年建成

ARCHITECT / 建筑师：Carlos Ott , Carlos Ponce de León Architects

Celebra Building, an emblematic double helicoid 8-storey office building in Zonamerica free zone, rises to celebrate the 2010 Technological Park´s 20th anniversary. Formed by two inverted sliced elliptical conic shapes, it achieves the required flexibility for office display, reaching the highest sets of demands attaining LEED GOLD certification. Celebra Building obtained in London the award for Best Office Building in the World at the International Property Awards, in association with Bloomberg.

塞勒贝拉大楼是佐纳美利加自由区中具有象征意义的双螺旋8层办公楼，为庆祝2010年技术园区成立20周年而建造。它由两个倒置的椭圆圆锥组成，可实现办公展示功能空间所需的灵活性，并获得美国LEED认证金级的最高要求。塞勒贝拉大楼与彭博社共同在伦敦获得了国际房地产奖颁发的“世界最佳办公楼”奖。

LEED

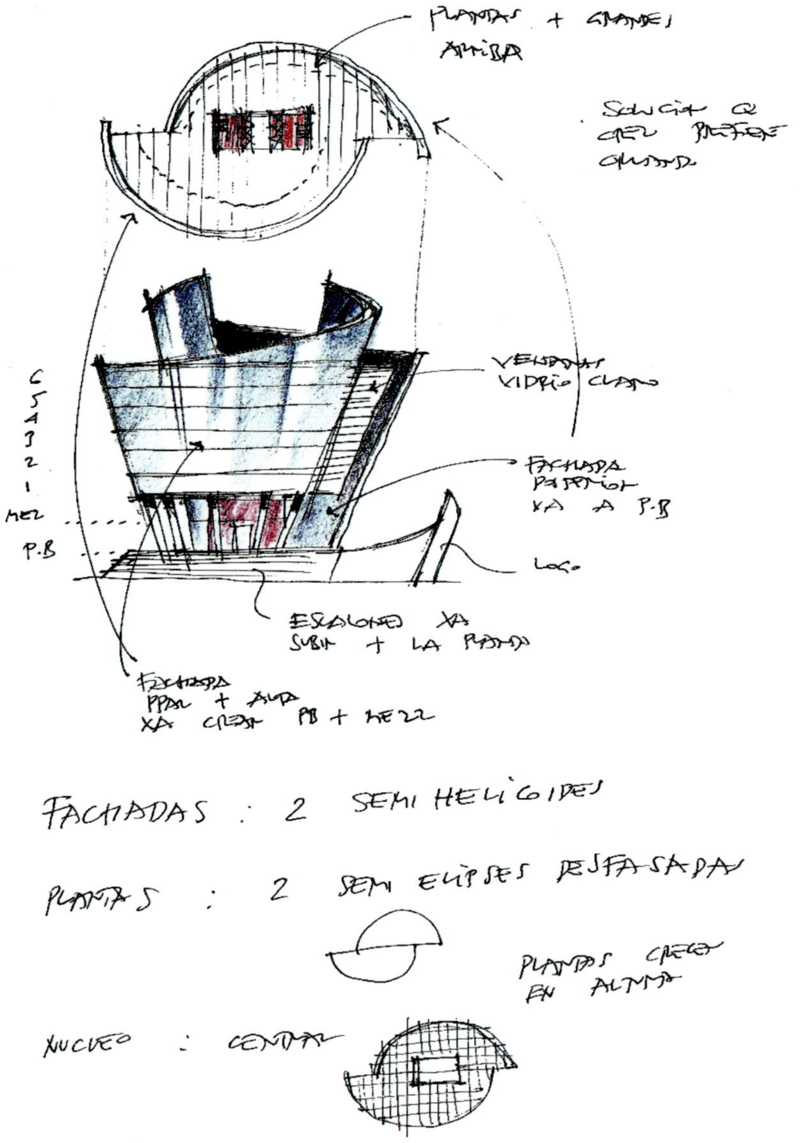

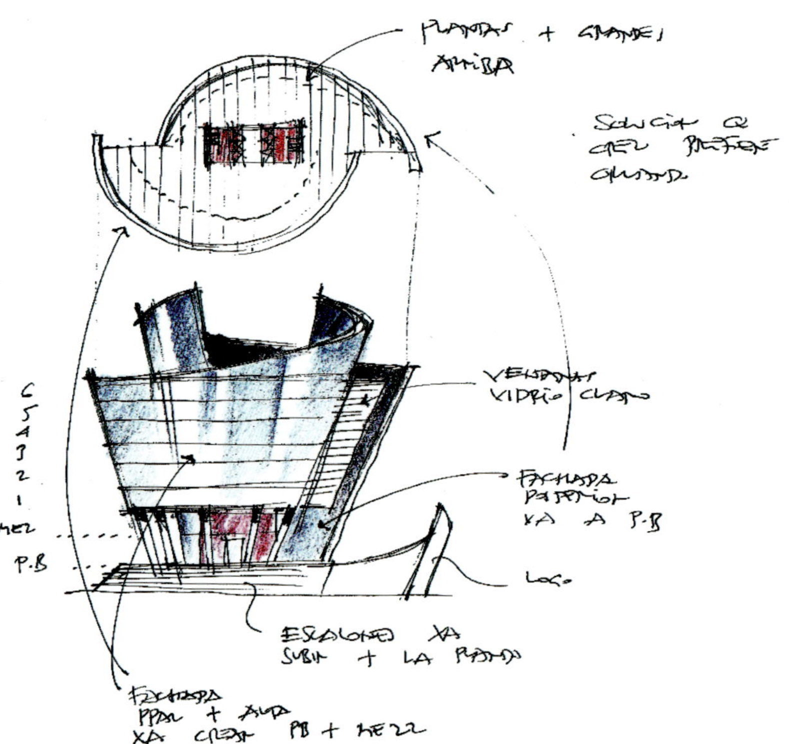

FACHADAS : 2 SEMI HELICOIDES

PLANTAS : 2 SEMI ELIPSES DESFASADAS

PLANTAS CRECEN EN ALTURA

NUCLEO : CENTRAL

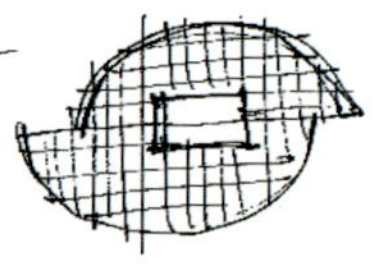

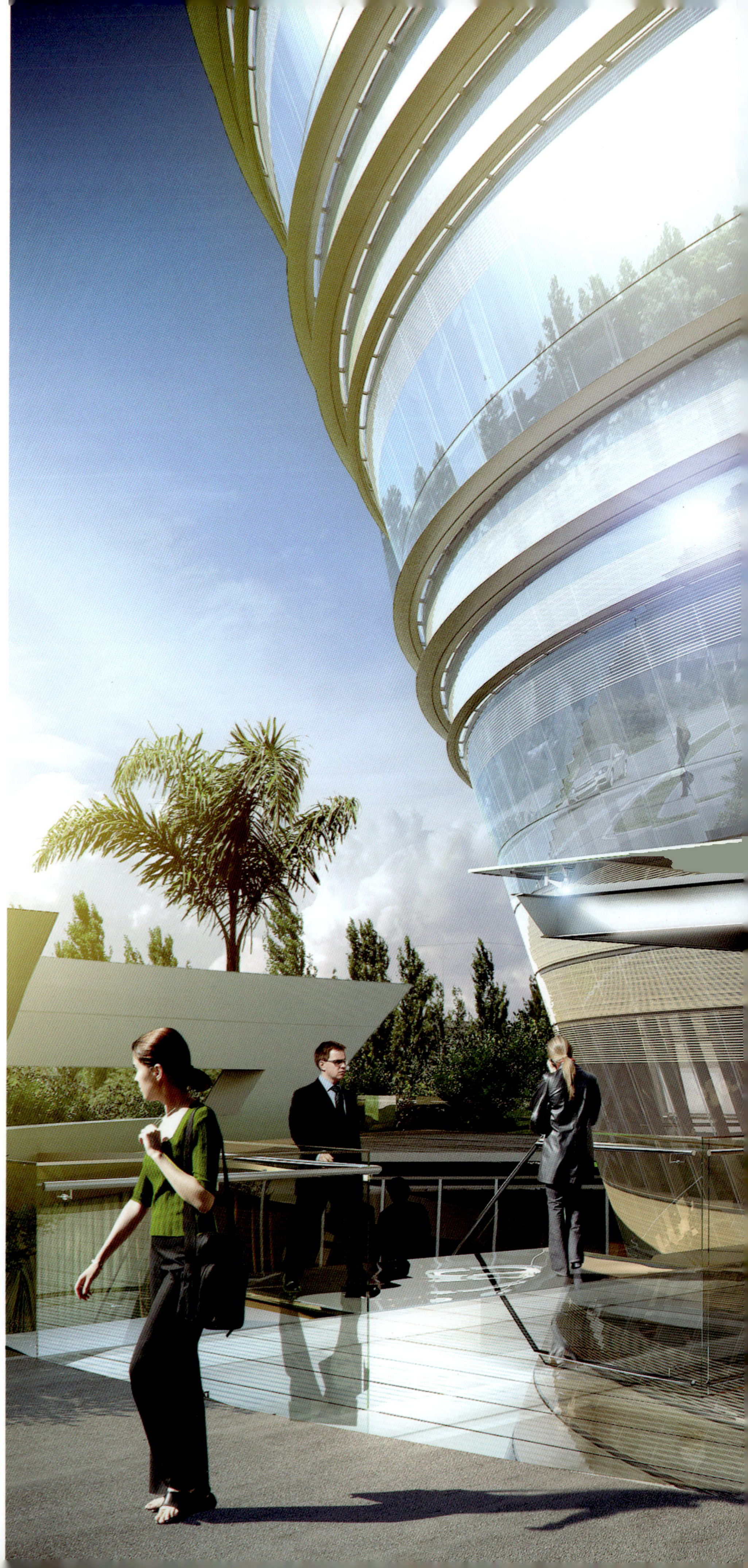

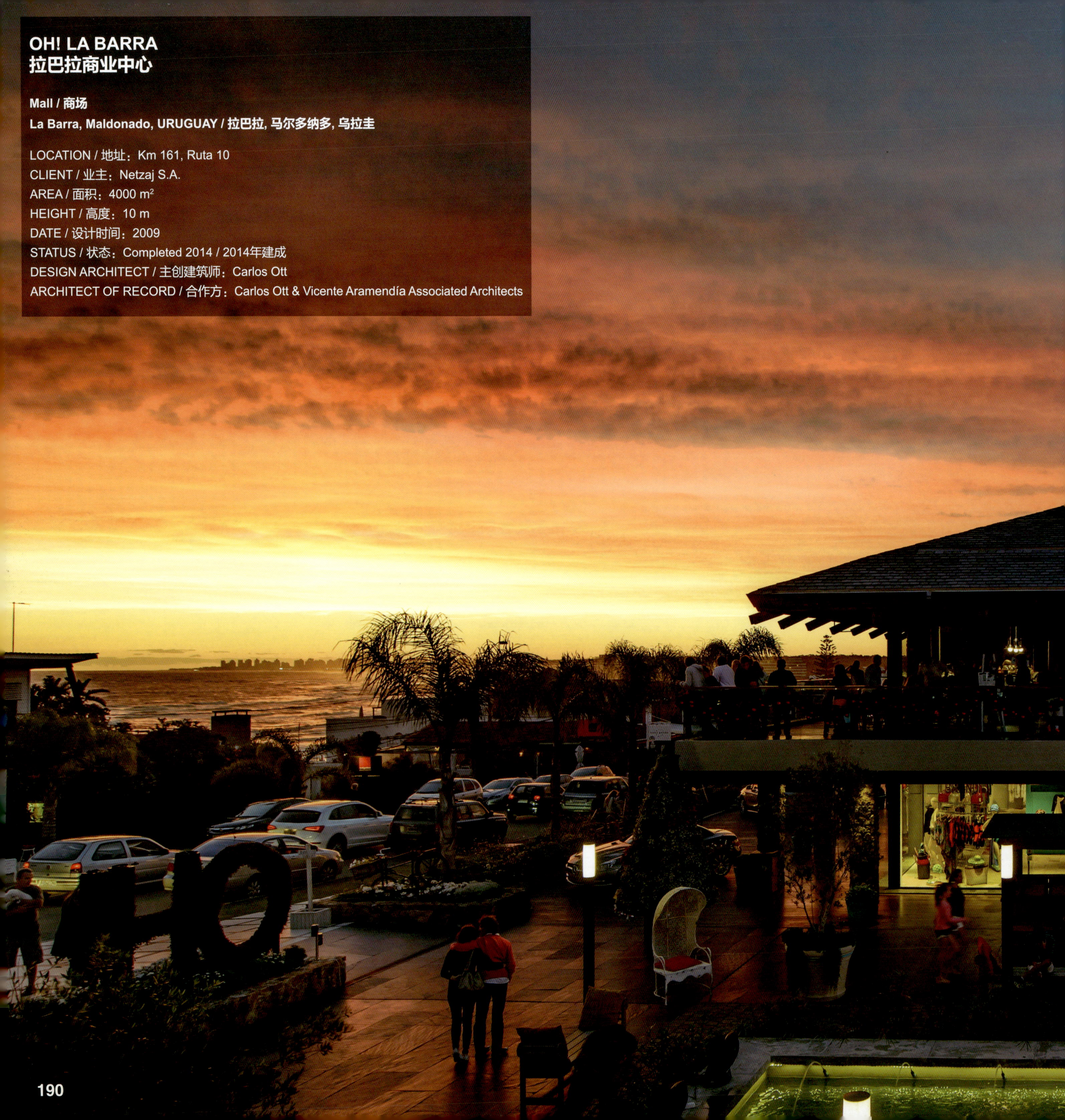

OH! LA BARRA
拉巴拉商业中心

Mall / 商场

La Barra, Maldonado, URUGUAY / 拉巴拉, 马尔多纳多, 乌拉圭

LOCATION / 地址：Km 161, Ruta 10
CLIENT / 业主：Netzaj S.A.
AREA / 面积：4000 m^2
HEIGHT / 高度：10 m
DATE / 设计时间：2009
STATUS / 状态：Completed 2014 / 2014年建成
DESIGN ARCHITECT / 主创建筑师：Carlos Ott
ARCHITECT OF RECORD / 合作方：Carlos Ott & Vicente Aramendía Associated Architects

The contemporary retail complex located in La Barra, an exclusive seaside area of Punta del Este, portrays the repetition of five 2-storey stone bungalows to create the antithesis of a concrete mall, allowing shops to have singular identities.

这座现代购物中心位于埃斯特角的豪华海滨地区。该商场由5个双层石制平房组成，使其中的商店独具特色。

HARBOUR TOWER
海湾公寓

Residential / 住宅

Buenos Aires, ARGENTINA / 布宜诺斯艾利斯, 阿根廷

AWARDS / 获奖：International Property Awards, 2015, Best International Residential High-Rise Architecture / 2015年度英国CNBC国际房地产大奖之全球最佳高层住宅建筑
LOCATION / 地址：Julieta Lanteri & Lola Mora St., Dique 1, Puerto Madero
CLIENT / 业主：GNV Group
AREA / 面积：50 500 m^2
HEIGHT / 高度：189.16 m, 52 storeys / 52层
UNITS / 户数：192
DATE / 设计时间：2014
STATUS / 状态：Under construction / 建设中
DESIGN ARCHITECT / 主创建筑师：Carlos Ott
ARCHITECT OF RECORD / 合作方：Urgel-Penedo-Urgell Architects

Located in Dique 1 of Madero Harbour in Buenos Aires, the 52-storey residential smart building with sustainable architecture Harbour Tower is the result of a glazed complex shape with panoramic views of the city and the river. The angled position of the building in relation to the plot broadens the views of the docks.

这座52层高的智能住宅建筑位于布宜诺斯艾利斯马德罗港的狄克1号，用可持续利用材质建成，表面由复杂玻璃面构成。建筑物与海港成一定角度，拓宽了码头的视野，因此人们在其中可欣赏到城市和河流的全景。

HARBOUR
TOWER

EZ TOWERS
EZ大厦

Corporate, Commercial / 企业总部, 商业

São Paulo, BRAZIL / 圣保罗, 巴西

AWARDS / 获奖 : LEED Gold Core & Shell, LEED Operations and Maintenance Certifications, Gold Winner (Office Category) / 美国绿色建筑委员会LEED标准（2009版）认证：主体结构及外围护系统金级（办公类）
FIABCI (International Real Estate Federation) World, Prix d´Excellence Awards 2018 / 2018年度世界卓越产业奖（世界不动产联盟颁发）

LOCATION / 地址：105 Arq Olavo Redig de Campos Ave., Chácara Santo Antônio

CLIENT / 业主：EZ TEC Empreendimentos e Participações S.A.

AREA / 面积：164 446 m^2

HEIGHT / 高度：146.7 m, 35 storeys / 35层

UNITS / 户数：224 offices / 224间办公室

DATE / 设计时间：2009

STATUS / 状态：Completed 2014 / 2014年建成

DESIGN ARCHITECT / 主创建筑师：Carlos Ott

ARCHITECT OF RECORD / 合作方：Silvio Iamamura

LANDSCAPE DESIGN / 景观建筑师：Benedito Abbud

INTERIOR DESIGN / 室内设计：Athié Wohnrath

Situated in the privileged central business area of Berrini-Chucri Zaidan Ave. in São Paulo, the 35-storey complex, with high technical, aesthetic, sustainable standards and LEED certification, includes two exclusive towers, a triple-A Class Corporate and an office tower. The towers have 26 office floors each.

这一高35层的建筑群位于圣保罗的中央商务区，具高科技、美学、可持续标准以及LEED认证，包括两座豪华办公大楼，一座AAA级企业办公楼和一座办公塔楼。每座大楼有26层办公楼。

BALSAM LAKE BOATHOUSE
鲍尔瑟姆湖船屋

Private Residence / 私人住宅

Kawartha Lakes, Ontario, CANADA / 卡瓦萨湖城, 安大略, 加拿大

LOCATION / 地址：Jasper Rd., Balsam Lake
CLIENT / 业主：J. Harlow
AREA / 面积：98 m^2
HEIGHT / 高度：7 m
DATE / 设计时间：2012
STATUS / 状态：Completed 2012 / 2012年建成
DESIGN ARCHITECT / 主创建筑师：Carlos Ott
ARCHITECT OF RECORD / 合作方：Replacement Design

Implanted in Kawarthas, an all-seasonal recreational park along the Trent-Severn Waterway, the Balsam Lake Harlow Boathouse, traditional yet modern, is made of glue laminated wood beams, slate roofs, double-glazed windows on black plastic covered aluminium frames, natural stone sloping walls, steel and glass. The cross gable roof, with equal-height horizontal perpendicular ridges, defines and octagonal layout; portraying the fusion of four skylights meeting in one pointGround floor that shelters two boats connects visually to the lakeside garden and the cedar deck with a typical South American barbecue "parrilla". Upper level, the refuge, is a den with glazed gables providing 270 degrees views, a fireplace, a library and the inner vision of the wooden roof.

鲍尔瑟姆湖船屋安置在沿特伦特-塞文水道的全季节休闲公园卡瓦萨斯之中，既传统又很现代，建筑由胶合木梁、石板屋顶，覆盖于黑色塑料铝框架上的双层玻璃窗，天然石头构筑的倾斜墙壁，以及钢和玻璃构成。交叉屋顶有等高的水平屋脊，形成八角形布局。四个天窗交融于一点。可以容纳两艘船的底楼在视觉上与湖边花园和雪松甲板相通，并带有典型的南美烧烤休闲风格。上层是避难层，是一个带有玻璃山墙的阁楼，提供270度的视野，有壁炉和藏书室，能看见船屋内景色。

SUNRISE TOWER
日出大厦

Residential / 住宅

Punta del Este, Maldonado, URUGUAY / 埃斯特角, 马尔多纳多, 乌拉圭

LOCATION / 地址：Pedragosa Sierra Ave., Parada 5, Playa Mansa
CLIENT / 业主：Jakaly S.A.
AREA / 面积：16 670 m^2
HEIGHT / 高度：63 m
UNITS / 户数：87
DATE / 设计时间：2010
STATUS / 状态：Completed 2014 / 2014年建成
DESIGN ARCHITECT / 主创建筑师：Carlos Ott, Silvia Menéndez Architect
ARCHITECT OF RECORD / 合作方：Maria José Delgado & Gustavo Lagomarsino Jakaly S.A.

Strategically located in Punta del Este, Sunrise Tower has views of Playa Mansa, the bay, the Peninsula harbour and the eastern Playa Brava, defined by smooth curvatures, continuous balconies and a grand rooftop canopy. The 87-unit tower with one level of underground parking includes four see-through units with 2 or 3 suites per level. At ground floor, the drop-off crosses from road to road, creating an entrance area linked to the double-height lobby, the indoor amenities with gym, spa and pool, and a tennis court implanted in the surrounding gardens. The rooftop provides three barbecue halls and solarium terraces.

埃斯特角的日出大厦有平滑的曲线、连续的阳台和夸张的屋顶挑檐，居民可以无尽欣赏曼萨海滩、海湾、半岛港口和布拉瓦东部地区的景色。大厦有87套住宅单元，1层地下停车场，4个透明单元，每层设有2个或3个套房。在地面层，落客区有小径连到主入口，该入口处与双层大堂相连，室内设施包括健身房、水疗中心和游泳池，周围花园中建有网球场。屋顶设有3个烧烤厅和日光浴露台。

SUMMER TOWER
夏日大厦

Residential / 住宅

Punta del Este, Maldonado, URUGUAY / 埃斯特角, 马尔多纳多, 乌拉圭

LOCATION / 地址：F. D. Roosevelt & M. Chiossi Ave., Parada 16
CLIENT / 业主：Drigey S.A.
AREA / 面积：10 960 m^2
HEIGHT / 高度：55 m, 22 storeys / 22层
UNITS / 户数：95
DATE / 设计时间：2010
STATUS / 状态：Completed 2016 / 2016年建成
DESIGN ARCHITECT / 主创建筑师：Carlos Ott, Silvia Menéndez Architect
ARCHITECT OF RECORD / 合作方：Maria José Delgado & Gustavo Lagomarsino Drigey S.A.

Overlooking Punta del Este´s Playa Mansa from the Peninsula´s harbour bay to Punta Ballena, the 22-storey Summer Tower rises unpretentiously, with reinforced vertical lines, crowned with arched planes.

夏日公寓从半岛的海港湾到蓬塔巴列纳一线俯瞰埃斯特角的曼萨海滩，这座22层高的夏日公寓毫不掩饰地耸立着，上面冠以拱形屋顶。

GARDENIA RESTAURANT
栀子花餐厅

Interior Design / 室内设计

Montevideo, URUGUAY / 蒙得维的亚，乌拉圭

LOCATION / 地址：World Trade Centre, Luis A. de Herrera & 26 de Marzo Ave., Buceo

CLIENT / 业主：Newco S.A.

AREA / 面积：340 m^2

DATE / 设计时间：2010

STATUS / 状态：Completed 2011 / 2011年建成

DESIGN ARCHITECT / 主创建筑师：Carlos Ott, Silvia Menéndez Architect

Located amidst prime restaurants that cater for the exclusive WTC offices plaza of Pocitos neighbourhood in Montevideo, Gardenia offers an exquisite menu within a refined modern minimalistic frame. Concrete floorings in grey shades, leather seats, a burgundy leather spiral reigning the helical staircase to the basement wine cellar and glazed facades facing both the WTC plaza and the yachting Buceo Harbour, give Gardenia a strong character.

为世贸中心广场附近办公楼提供餐饮的栀子花餐厅位于蒙得维的亚附近的波西托斯，餐厅设计现代、简约精致，提供的菜品精美独特。灰色的混凝土地面，皮革座椅，勃艮第皮革装饰的螺旋楼梯通往地下酒窖，以及面向世贸中心广场和游艇码头的玻璃幕墙，赋予了栀子花餐厅强烈的个性。

CORP LAGOS
拉各斯集团办公大楼

Commercial / 商业
Ciudad de la Costa, Canelones, URUGUAY / 海岸城, 卡内隆市, 乌拉圭

LOCATION / 地址：De las Américas Ave. & Canelones St., Paso Carrasco
CLIENT / 业主：Ciromare S.A.
AREA / 面积：1293 m^2
HEIGHT / 高度：15m, 6 storeys / 6层
UNITS / 户数：20
DATE / 设计时间：2011
STATUS / 状态：Completed 2014 / 2014年建成
DESIGN ARCHITECT / 主创建筑师：Carlos Ott, Silvia Menéndez Associated Architect
ARCHITECT OF RECORD / 合作方：Robert Corrales Ciromare S.A.

CORP LAGOS commercial and office building set on a privileged location within a green residential area, very close to Montevideo access and key business services as Carrasco International Airport , Zonamerica Free Zone, and Ciudad de la Costa Civic Centre.

拉各斯集团办公大楼位于绿色居住区的优越位置，靠近蒙得维的亚，其主要商业服务包括卡拉斯科国际机场、佐纳美利加自由区和哥斯达黎加市民中心。

The International Convention and Exhibition Centre has a strategic location in a 12 000 m² in the renowned sea resort of South America, Punta del Este.

国际会展览中心地理位置优越，位于南美著名的海滨胜地埃斯特角，占地面积12 000 m²。

PUNTA DEL ESTE CONVENTION & EXHIBITION CENTRE
埃斯特角会展中心

Convention Centre / 会展中心
Punta del Este, Maldonado, URUGUAY / 埃斯特角, 马尔多纳多, 乌拉圭

LOCATION / 地址：Pedragosa Sierra & Aparicio Saravia Ave., Jagüel
CLIENT / 业主：Maldonado Municipality
AREA / 面积：18 000 m²
HEIGHT / 高度：12 m
DATE / 设计时间：2010
STATUS / 状态：Completed 2016 / 2016年建成
DESIGN ARCHITECT / 主创建筑师：Carlos Ott, Bilcoy S.A.
ARCHITECT OF RECORD / 合作方：Abengoa Teyma

ARTOWER
阿尔特公寓

Residential / 住宅

Punta del Este, Maldonado, URUGUAY / 埃斯特角, 马尔多纳多, 乌拉圭

LOCATION / 地址：Chiverta Ave. & Francisco Salazar Ave., Parada 4
CLIENT / 业主：Mesit S.A.
AREA / 面积：24 000 m^2
HEIGHT / 高度：73 m, 23 storeys / 23层
DATE / 设计时间：2010
STATUS / 状态：Completed 2020 / 2020年建成
DESIGN ARCHITECT / 主创建筑师：Carlos Ott
ARCHITECT OF RECORD / 合作方：Silvia Menéndez Architect, Atijas Casal, Pablo Atchugarry (artist)

In the centre of the Peninsula of Punta del Este, close to both main beaches, Mansa and Brava, the 23-storey residential tower faces the corner with a circular floor plan of 73 metres of height. It has two basements for vehicle parking and general building services. The double-height circular reception lobby designed around the core contains the elevators lobby and exit doors to exterior spaces, connecting with two independent 2-storey buildings where the main amenities, gym, spa and indoor pool are located.

这座23层高的住宅大楼位于埃斯特角半岛的中心，靠近两个主要海滩，曼萨海滩和布拉瓦海滩，73米高的建筑拥有圆形的平面，以尽可能面向海角。大楼有两个地下室，用于停车场和常规物业服务。围绕核心筒设计的双层通高圆形接待大厅包含电梯大厅和通往外部空间的出入口，且与两座独立的两层建筑相连，这两座双层建筑中有大楼配套设施、健身房、水疗中心和室内游泳池。

LAB LAS AMÉRICAS BUREAU
LAB办公大楼

Commercial, Retail / 商业, 零售

Ciudad de la Costa, Canelones, URUGUAY / 海岸城, 卡内隆市, 乌拉圭

LOCATION / 地址：Las Américas Ave. & Queguay St., Miramar Park
CLIENT / 业主：Ciromare S.A.
AREA / 面积：1650 m^2
HEIGHT / 高度：8 storeys / 8层
UNITS / 户数：17 offices / 17间办公室
DATE / 设计时间：2010
STATUS / 状态：Completed 2018 / 2018年建成
DESIGN ARCHITECT / 主创建筑师：Carlos Ott , Silvia Menéndez Associated Architect
ARCHITECT OF RECORD / 合作方：Robert Corrales-Ciromare S.A.

LAB Las Americas Bureau designed for rental offices is strategically located next to Carrasco International Airport, Zonamerica Free Zone and Ciudad de la Costa Civic Centre. Visible and iconic, the 8-storey angular layout takes advantage of natural light and views of the surrounding lakes and Roosevelt Park. At grade level, an enclosed plaza integrates the off-street parking and the landscaped drop-off areas. The double hermetic glazing DVH facades limit solar radiation and brightness by alternating fixed and moving panels giving natural ventilation to all offices.

LAB办公大楼专门为租赁式办公室而设，地理位置优越，毗邻卡拉斯科国际机场，佐纳美利加免税区和哥斯达黎加市民中心。8层高的角形布局清晰可见，具有标志性，充分利用了自然光线，可欣赏周围湖泊和罗斯福公园的景致。在地面层，一个内向型的广场将道外停车场和景观区融为一体。双层全密封DVH玻璃幕墙通过可移动、可替换面板增强了自然通风，并减少了太阳辐射和直射。

SHENNONG GRAND THEATRE
神农大剧院

Cultural / 文化建筑

Zhuzhou, Hunan, CHINA / 株洲, 湖南, 中国

LOCATION / 地址：Shennong Ave. & Taishan Rd., Tianyuan District
CLIENT / 业主：Urban Construction Headquaters of Shennong City
AREA / 面积：44 000 m^2
HEIGHT / 高度：87 m
DATE / 设计时间：2010
STATUS / 状态：Completed 2018 / 2018年建成
DESIGN ARCHITECT / 主创建筑师：Carlos Ott

Shennong Grand Theater is located in the southwest corner of the core area of Shennong City, Zhuzhou. It is an important node project in the core area of Shennong City, an iconic cultural facility in Zhuzhou City, and an important platform and window for displaying local culture. The Grand Theater consists of a 1400-seat theater, a 400-seat experimental theater and supporting auxiliary rooms, with a total construction area of 45 991 m^2. It takes "mountain high and man as the peak" as its theme, showing the brave, free, strong and enterprising spirit of Shennong, as well as the spirit of self-improvement and enthusiasm for the holy fire, thus expressing respect and memory of Emperor Shennong Yan. The architectural shape of the Grand Theater presents a spiral upward trend of the character "human", which resembles peaks uplifted on the ground, which is very iconic and memorial.

神农大剧院位于株洲神农城核心区西南角，是神农城核心区重要的节点工程、株洲市的标志性文化设施、展示本土文化的重要平台与窗口。神农大剧院由拥有1400个座位的剧院、400个座位的实验剧场及配套辅助用房组成，总建筑面积45 991 m^2。大剧院以"山高人为峰"为主题，展现勇敢自由、坚强进取的神农精神，以及自强不息、热情奔放的圣火精神，从而表达对神农炎帝的敬仰与纪念。大剧院建筑造型呈现"人"字形的螺旋上升态势，似平地拔起的峰峦，极具标志性与纪念性。

金龙大厦

HANGZHOU LOW-CARBON TECHNOLOGY MUSEUM
杭州低碳科技馆

Cultural / 文化建筑

Hangzhou, Zhejiang, CHINA / 杭州, 浙江, 中国

LOCATION / 地址：No.1888 Jianghan Rd., Bingjiang, Hangzhou
CLIENT / 业主：Hangzhou Low-Carbon Technology Museum
AREA / 面积：105 000 m^2
DATE / 设计时间：2007
STATUS / 状态：Completed 2012 / 2012年建成
DESIGN ARCHITECT / 主创建筑师：Carlos Ott, PPA Architects
ARCHITECT OF RECORD / 合作方：Petroff Partnership Architects , China Aviation Planning and Design Institute

Hangzhou Low-Carbon Science and Technology Museum is the world´s first large scale museum with a low-carbon theme.It is a public science institution that integrates low-carbon technology popularization, green buildiing display and low-carbon academic exchanges and dissemination. Educationally, it is the "second classroom", especially for young people, to learn about low-carbon life, low-carbon cities and low carbon economy.

杭州低碳科技馆是全球第一家以低碳为主题的大型科技馆；是集低碳科技普及、绿色建筑展示、低碳学术交流和低碳信息传播等职能为一体的公益性科普教育机构；是公众，特别是青少年了解低碳生活、低碳城市、低碳经济的"第二课堂"。

UADE CAMPUS COSTA ARGENTINA
阿根廷商业大学皮纳马校区

Education / 教育

Pinamar, Buenos Aires, ARGENTINA / 布宜诺斯艾利斯, 阿根廷

LOCATION / 地址：776 Intermédanos Sur Ave.
CLIENT / 业主：Universidad Argentina de la Empresa
AREA / 面积：5400 m^2
HEIGHT / 高度：11.5 m, 2 storeys / 2层
DATE / 设计时间：2010
STATUS / 状态：Completed 2012 / 2012年建成
DESIGN ARCHITECT / 主创建筑师：Carlos Ott
ARCHITECT OF RECORD / 合作方：Oficina Urbana S.A., Roberto Converti Architect

Aiming to generate a regional educational centre, the 2-storey complex for the University of Hospitality and Gastronomy centre, implanted within the existing pine forests of the exclusive sea resort Pinamar, includes laboratories, an auditorium and residential areas. Articulated volumes with covered walkways surrounding a courtyard,interconnect classrooms,library, amphitheatre, food elaboration and the hotel area.

为了打造一个区域性的教育中心，酒店与餐饮管理学院的两层综合楼群规划在了海滨度假胜地皮纳马尔现存的松树林中，包括实验室、礼堂和居住区。围绕着院子的廊道将教室、图书馆、圆形剧场、食品加工和酒店区域互相连接起来。

APOGEE BEACH
阿普奇海滩公寓

Residential / 住宅

Hollywood, Florida, USA / 好莱坞, 佛罗里达, 美国

LOCATION / 地址：3951 S. Ocean Drive, Hollywood Beach
CLIENT / 业主：Related Group
AREA / 面积：15 400 m^2
HEIGHT / 高度：22 storeys / 22层
UNITS / 户数：49
DATE / 设计时间：2010
STATUS / 状态：Completed 2013 / 2013年建成
DESIGN ARCHITECT / 主创建筑师：Carlos Ott, CFE Architects
ARCHITECT OF RECORD / 合作方：CFE Architects

Developing along 73 metres of Hollywood Beach, the 22-storey Apogee Beach curved complex has views over the Atlantic, Port of Miami, and Biscayne Bay, with 49 suites of two, three and four-bedroom residences, and up to 3.35-metre wide terraces, private cabanas and an oceanfront pool.

阿普奇海滩公寓呈曲线形，沿着73米的好莱坞海滩，高22层，可同时欣赏大西洋、迈阿密港和比斯坎湾的美景。49套公寓配置有两居室、三居室和四居室住宅，以及最大宽度达3.35米的私人露台，遮阳小屋和海滨游泳池。

FORUM BUCEO HARBOUR
FORUM海湾公寓

Residential / 住宅

Montevideo, URUGUAY / 蒙得维的亚, 乌拉圭

LOCATION / 地址：Rambla Armenia & Rambla República Federal de Alemania, Buceo

CLIENT / 业主：FDB S.A.

AREA / 面积：60 000 m^2

HEIGHT / 高度：9 storeys / 9层

UNITS / 户数：270

DATE / 设计时间：2012

STATUS / 状态：Completed 2019 / 2019年建成

DESIGN ARCHITECT / 主创建筑师：Carlos Ott, Carlos Ponce de León Architects

PROSCENIUM AT ROCKWELL
普罗塞尼欧公寓

Residential, Commercial, Cultural / 住宅, 商业, 文化建筑
Makati City, Manila, PHILIPPINES / 马卡蒂, 马尼拉, 菲律宾

LOCATION / 地址：J.P. Rizal Ave. & Estrella St., Rockwell Centre
CLIENT / 业主：Rockwell Land Corporation
AREA / 面积：280 100 m^2
HEIGHT / 高度：191 m, 47 to 60 storeys / 47～60层
UNITS / 户数：1414
DATE / 设计时间：2011
STATUS / 状态：Completed 2018－2019 / 2018－2019年建成
DESIGN ARCHITECT / 主创建筑师：Carlos Ott
ARCHITECT OF RECORD / 合作方：Pimentel Rodriguez Simbulan & Partners Architects

The Proscenium, whose name is inspired by the performing arts theatre within the 36 000 m² complex property of Rockwell Centre in Makati, is a residential and commercial condominium, comprised by five 47 to 60-storey residential towers, including the main tower with a distinct, iconic design to complete a unique skyline.

普罗塞尼欧公寓是一个住宅和商业综合建筑，由5座47层至60层的住宅大楼组成，其中包括具有独特风格的主楼。该公寓楼的名称受到马卡迪罗克韦尔中心3.6万平方米综合大楼内的表演艺术剧院的启发而得名。其标志性的设计为大楼形成的独特的天际线。

BIKINI BEACH
比基尼海滩公寓

Residential / 住宅

Manantiales, Maldonado, URUGUAY / 马南蒂亚莱斯，马尔多纳多, 乌拉圭

LOCATION / 地址：Ruta 10 & Solanas St.
CLIENT / 业主：Bakuly S.A.
AREA / 面积：2094 m^2
HEIGHT / 高度：12 m, 4 storeys / 4层
UNITS / 户数：17
DATE / 设计时间：2011
STATUS / 状态：Completed 2015 / 2015年建成
DESIGN ARCHITECT / 主创建筑师：Carlos Ott
ARCHITECT OF RECORD / 合作方：Matias Casaux

Located on a high site overlooking Manantiales´ beaches in the Uruguayan exclusive sea resort Punta del Este, the arched 4-storey, 17-apartment residential building of 1 to 3 bedrooms, includes an indoor-outdoor pool, gym and barbecue. Composed by continuous balconies with clear glass handrails, fully glazed units, stone fins, white aluminium canopies, a circular drop-off surrounding a water feature and 22 parking stalls.

这座弧形的住宅楼有4层楼高，共17户，位于乌拉圭独特的海滨度假胜地埃斯特角，俯瞰马南蒂亚莱斯的海滩，每个单元有1至3间卧室，设有一个半露天游泳池，还有健身房和烧烤设施。外立面上的连续阳台由玻璃栏板相隔，全透明厅，石质嵌条，白色铝质挑檐，围绕水景的圆形落客区和22个停车位。

VIRGEN DE LAS ARENAS CHAPEL
圣母教堂

Religious / 宗教建筑

José Ignacio, Maldonado, URUGUAY / 何塞·伊格纳西奥, 马尔多纳多, 乌拉圭

LOCATION / 地址：Laguna Escondida, Arenas de José Ignacio
CLIENT / 业主：Laguna Escondida Beach & Lagoon Resort
AREA / 面积：200 m^2
HEIGHT / 高度：14 m
DATE / 设计时间：2016
STATUS / 状态：Completed 2016 / 2016年建成
DESIGN ARCHITECT / 主创建筑师：Carlos Ott

The chapel is located within Laguna Escondida Beach & Lagoon, an exclusive gated summer resort in José Ignacio; the top seaside location on the Atlantic Ocean in Punta del Este, Uruguay.The masterplan includes beach service, a boutique hotel in the bay, ocean views surrounded by the forest, with countryside peace and lagoon serenity.

教堂位于拉古纳·埃斯孔迪达海滩和泻湖内，好似与自然融合而来的雕塑。两只手合在一起的祷告台由黑色钢结构、木板和混凝土组建，构筑了教堂的直纹表面。屋顶通过敞开的天窗将空间照亮，而祭坛则将泻湖的景色尽收眼底。

ABITAB HEADQUARTERS
ABITAB银行总部大楼

Corporate / 企业总部

Montevideo, URUGUAY / 蒙得维的亚, 乌拉圭

LOCATION / 地址：Fernández Crespo Ave. & Batoví St., Cordon

CLIENT / 业主：ABITAB

AREA / 面积：7511 m^2

HEIGHT / 高度：19 m, 6 storeys / 6层

DATE / 设计时间：2012

STATUS / 状态：Completed 2016 / 2016年建成

DESIGN ARCHITECT / 主创建筑师：Carlos Ott, Carlos Ponce de León Architects

Located in a central area of Montevideo, the fully glazed 6-storey functional complex, with basement and ground floor parking, contains a financial and service institution (Abitab´s headquarters) office building. Geometrical opposition of fully glazed volumes define a conceptual clarity, while incorporating technological devices, information and communication technology, thermal conditioning and natural lighting. Office layouts are flexible, with maximum natural lighting and ventilation, energy saving and carbon dioxide emission reduction, thus comprising the maximum of natural resources.

通体玻璃的ABITAB银行总部大楼位于蒙得维的亚的中心地区，是金融服务机构办公大楼，带有地下室和地下停车场。在融合了技术设备，信息和通信技术，热调节和自然采光的同时，办公室布局灵活，最大限度地利用自然采光和通风，既节能和减少二氧化碳排放量，又最大限度地利用了自然资源。

GREENS
格林斯公寓

Residential / 住宅

Ciudad de la Costa, Canelones, URUGUAY /

海岸城, 卡内隆市, 乌拉圭

LOCATION / 地址：Batovi & Manuel Oribe St., Barra de Carrasco

CLIENT / 业主：Jardines Urbanos Nidecar S.A.

AREA / 面积：8500 m^2

HEIGHT / 高度：2 storgs / 2层

UNITS / 户数：64

DATE / 设计时间：2011

STATUS / 状态：Completed 2018 / 2018年建成

DESIGN ARCHITECT / 主创建筑师：Carlos Ott, Silvia Menéndez Associated Architect

ARCHITECT OF RECORD / 合作方：Luis Croci

Sixteen 2-storey single family residences (123 m^2 each) with individual parking, surrounding a green park with amenities in a clubhouse in a residential enclosed neighbourhood.

该公寓拥有16栋两层楼的独立车位单户住宅（每间123 m^2），环绕着一个绿色公园，形成整体的、设施便利的、完整的邻里中心。

1100 MILLECENTO
1100 MILLECENTO公寓

Residential & Commercial / 住宅, 商业

Miami, Florida, USA / 迈阿密, 佛罗里达, 美国

LOCATION / 地址：1100 South Miami Ave., Brickell
CLIENT / 业主：Related Group
AREA / 面积：61 370 m^2
HEIGHT / 高度：127 m, 42 storeys / 42层
UNITS / 户数：382
DATE / 设计时间：2012
STATUS / 状态：Completed 2015 / 2015年建成
DESIGN ARCHITECT / 主创建筑师：Carlos Ott
ARCHITECT OF RECORD / 合作方：CFE Architects
INTERIOR DESIGN / 室内设计：Pininfarina Home Design

42-storey high rise residential tower in the centre of Brickell, Miami´s business area. Variety of units mix (ranging from 54 to 118 m^2) including studios, one and two bedrooms, and three commercial units.

该公寓位于迈阿密商业区布里克尔中心，有42层，有各种户型（54～118 m^2），如工作室、一房和两房公寓，以及3个商业单元。

EZ MARK
EZ 马克商务中心

Commercial / 商业

São Paulo, BRAZIL / 圣保罗, 巴西

LOCATION / 地址：Dr. Tirso Martins St. & Domingos de Morais Ave.
CLIENT / 业主：EZ TEC Empreendimentos e Participações S.A.
AREA / 面积：21 000 m^2
HEIGHT / 高度：27.5 m, 7 storeys / 7层
DATE / 设计时间：2013
STATUS / 状态：Completed 2017 / 2017年建成
DESIGN ARCHITECT / 主创建筑师：Carlos Ott
ARCHITECT OF RECORD / 合作方：Silvio Iamamura
INTERIOR DESIGN / 室内设计：Priscilla Zarzur

Low rise 7-storey office building with both flexible office floors going from 50 to 1500 m^2 areas and small studios with a restaurant is in the centre in one of São Paulo´s major business area. The structure provides a different type of office space with a major building but with a low height. The symmetrical facade has at the centre a small cafe that acts as the main entrance to the complex. Inclined glass facades provide a dynamic appearance to the composition.

这是一幢位于圣保罗重要中心商务区的办公楼，共有7层，既有灵活的办公楼层，面积从50 m^2到1500 m^2不等，还包含餐厅及小型工作室。该建筑提供了另一类办公空间，作为一幢重要建筑，却有相对低矮的楼高。对称立面的中心处设有一个小咖啡馆，是该建筑群的主要入口。倾斜的玻璃幕墙为建筑群提供了动感的外观。

EZ MARK
VILA MARIANA •

SAGE
赛格海滩公寓

Residential / 住宅

Hollywood, Florida, USA / 好莱坞, 佛罗里达, 美国

LOCATION / 地址：2101, 2205 S. Surf Rd.
CLIENT / 业主：Property Markets Group
AREA / 面积：8900 m^2
HEIGHT / 高度：5 storeys / 5层
UNITS / 户数：24
DATE / 设计时间：2010
STATUS / 状态：Completed 2016 / 2016年建成
DESIGN ARCHITECT / 主创建筑师：Carlos Ott
ARCHITECT OF RECORD / 合作方：CFE Architects

5-storey boutique condominium with 24 two-and-three bedroom units is on the beach of one of southern Florida's most important locations. The roof top units include private pools and open air terraces.The project provides a more intimate atmosphere to the surrounding tall and large high rise towers, creating a direct dialog between the units and the white sand beach.

这幢5层精品综合公寓位于佛罗里达州南部最重要的一个海滩上，共有24个双卧室和三卧室单元。屋顶单元具有私人游泳池和开放露台。该公寓与周围的高层和大型公寓相适应，与白色沙滩相融合。

ECHO AVENTURA
ECHO 阿文图拉公寓

Residential / 住宅

Aventura, Florida, USA / 阿文图拉, 佛罗里达, 美国

LOCATION / 地址：3250 NE 188th St.
CLIENT / 业主：Property Markets Group
AREA / 面积：89 720 m^2
HEIGHT / 高度：41 m, 11 storeys / 11层
UNITS / 户数：190
DATE / 设计时间：2012
STATUS / 状态：Completed 2015 / 2015年建成
DESIGN ARCHITECT / 主创建筑师：Carlos Ott
ARCHITECT OF RECORD / 合作方：CFE Architects
INTERIOR DESIGN / 室内设计：Yabu Pushelberg

Low rise 11-storey prime residential condominium has 172 units of different sizes and 18 penthouses, with marina docks along Florida Intracoastal Waterway, including large infinity edge swimming pool and major fitness centre. This “low rise” structure provides large units with major terraces incorporating barbeques to take advantage of the direct views to the waterway, the ocean, downtown Miami and sunsets.In order to reduce apparent length, the building is split into a concave and a convex crescent shaped forms, in the centre of which is the main access and amenities.

该幢公寓共有11层，包含172个不同大小的单元和18个顶层单元，沿佛罗里达内河航道一侧设有码头，包括大型无边泳池和大型健身中心。这种“低层”建筑为大户型住宅提供了带有烧烤的露台，同时可直接欣赏水道、海洋、迈阿密市区景观和日落的景致。为了减少外观长度，建筑物被分为凹形和凸形，其中间是主要入口和住宅配套设施。

OCAMPO RESIDENCE
奥坎波私宅

Private Residence / 私人住宅

José Ignacio, Maldonado, URUGUAY / 何塞·伊格纳西奥, 马尔多纳多, 乌拉圭

LOCATION / 地址：Ruta 10, Arenas de José Ignacio
CLIENT / 业主：Emilio Ocampo
AREA / 面积：580 m^2
HEIGHT / 高度：3 storeys / 3层
DATE / 设计时间：2012
STATUS / 状态：Completed 2013 / 2013年建成
DESIGN ARCHITECT / 主创建筑师：Carlos Ott
ARCHITECT OF RECORD / 合作方：Matías Casaux

This 3-storey private residence includes a separate guest unit, a living room with a garage and glass floor looking down at the swimming pool below. It is Located in one of Uruguay´s top seaside private resort.

该建筑位于乌拉圭顶级的海滨私人度假胜地之一。这个三层私人住宅，包括一个独立的宾客单元和一个可以向下看到泳池的带车库的装有玻璃地板的客厅。

ECHO BRICKELL
ECHO 布里克尔公寓

Residential / 住宅

Miami, Florida, USA / 迈阿密, 佛罗里达, 美国

LOCATION / 地址：1451 Brickell Ave.

CLIENT / 业主：Property Markets Group

AREA / 面积：39 897 m^2

HEIGHT / 高度：230 m, 57 storeys / 57层

UNITS / 户数：157

DATE / 设计时间：2013

STATUS / 状态：Completed 2017 / 2017年建成

DESIGN ARCHITECT / 主创建筑师：Carlos Ott

ARCHITECTS OF RECORD / 合作方：CFE Architects

INTERIOR DESIGN / 室内设计：Carlos Ott + Yoo

It is a 6-storey residential tower on Brickell Ave. with 1 to 4 bedrooms condos and penthouses, the main street in the financial district of Miami. Two storeys amenities centre is in the centre of the building including swimming pool with open views to the ocean. The tower provides fully robotic parking with several storeys.A sinusoidal profile creates a distinct image in the Brickell corridor of Miami surrounded by high rise buildings.Its views from north and south bound traffic makes an immediately recognizable iconic structure.

该公寓位于迈阿密金融区的主要街道布里克尔大街，共有60层，设有1至4间卧室的公寓和顶层公寓。两层楼的设备中心位于大楼的中部，这里有可欣赏到大海美景的游泳池。大楼有完全机械化的多层停车场。在迈阿密布里克尔大街耸立的高楼中，这幢曲线形的大楼非常醒目。它在南北向交通的视野中成为一个高辨识度的标志性建筑。

MUSE
缪斯女神公寓

Residential / 住宅

Sunny Isles Beach, Florida, USA / 阳光岛海滩, 佛罗里达, 美国

LOCATION / 地址：17141 Collins Ave., Sunny Isles, Florida
CLIENT / 业主：Property Markets Group, S2 Developments LLC
AREA / 面积：37 900 m^2
HEIGHT / 高度：194 m, 47 storeys / 47层
UNITS / 户数：68
DATE / 设计时间：2016
STATUS / 状态：Completed 2019 / 2019年建成
DESIGN ARCHITECT / 主创建筑师：Carlos Ott
ARCHITECT OF RECORD / 合作方：Sieger Suarez Architects
INTERIOR DESIGN / 室内设计：Antrobus + Ramirez

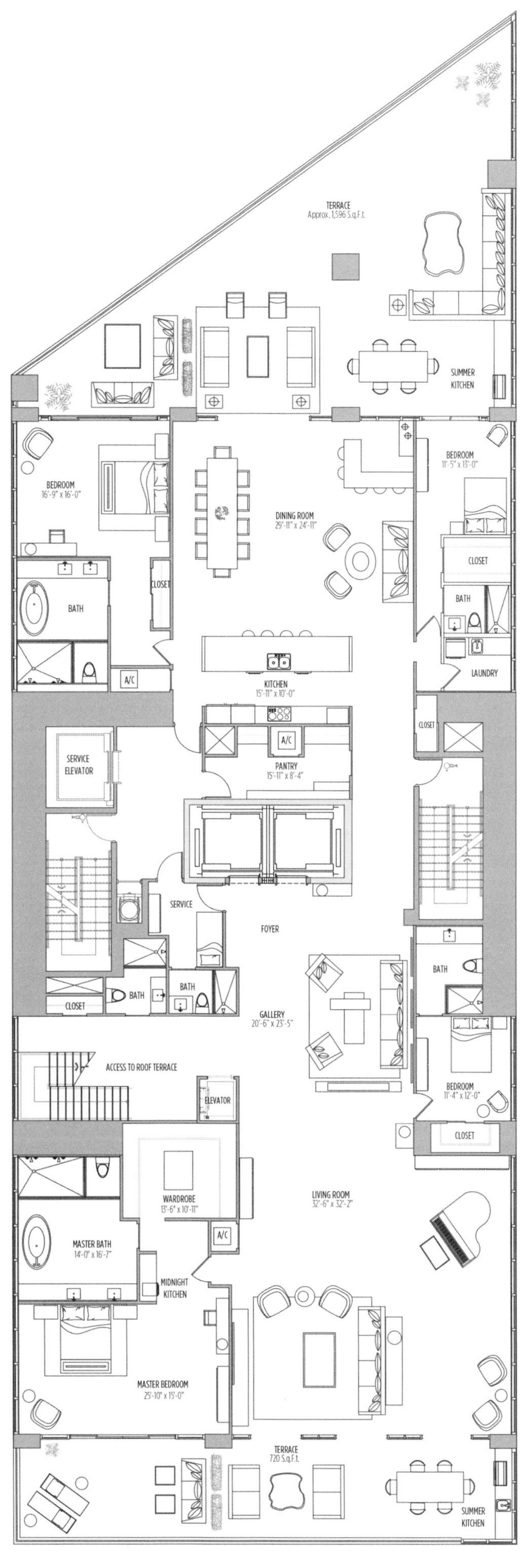
TERRACE
Approx. 1,596 S.q.F.t.
SUMMER
KITCHEN
BEDROOM
16'-9" x 16'-0"
DINING ROOM
29'-11" x 24'-11"
BEDROOM
11'-5" x 13'-0"
CLOSET
CLOSET
BATH
BATH
A/C
KITCHEN
15'-11" x 10'-0"
LAUNDRY
CLOSET
SERVICE
ELEVATOR
A/C
PANTRY
15'-11" x 8'-4"
SERVICE
FOYER
BATH
BATH
BATH
CLOSET
GALLERY
20'-6" x 23'-5"
ACCESS TO ROOF TERRACE
ELEVATOR
BEDROOM
11'-4" x 12'-0"
CLOSET
WARDROBE
13'-6" x 10'-11"
LIVING ROOM
32'-6" x 32'-2"
A/C
MASTER BATH
14'-0" x 16'-7"
MIDNIGHT
KITCHEN
MASTER BEDROOM
25'-10" x 15'-0"
TERRACE
720 S.q.F.t.
SUMMER
KITCHEN

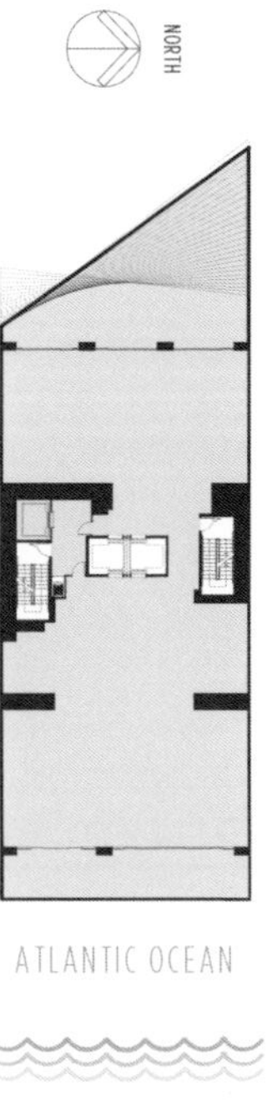
NORTH
ATLANTIC OCEAN

It is a 47-storey prime residential tower on Collins Avenue on Sunny Isles Beach with one or maximum two units per floor, fully robotic parking, swimming pool at beach level and restaurant at ground floor. Since it is on a very narrow site facing the ocean and Collins Ave., the solution was to design a 20-storey robotic parking garage on the west side with 8 cars per level, 4 at each side of the two car elevators, since tall structures, are planned across the street.

这幢47层的高端住宅楼位于阳光岛海滩上的柯林斯大道，每层有一个或最多两个单元，全机械化停车场，海滩层有游泳池，地面层有餐厅。选址位于一个面向海洋和柯林斯大道的狭窄区域，解决方案是在西侧设计一个20层高的机械停车库，每层可停8辆车，在两部汽车电梯的每一侧各停4辆。

ALMA SUR
阿尔玛公寓

Residential / 住宅

Montevideo, URUGUAY / 蒙得维的亚, 乌拉圭

LOCATION / 地址：Andes & Maldonado St., Downtown Montevideo
CLIENT / 业主：New Design S.A.
AREA / 面积：8180 m^2
HEIGHT / 高度：27 m, 12 storeys / 12层
UNITS / 户数：78
DATE / 设计时间：2013
STATUS / 状态：Completed 2017 / 2017年建成
DESIGN ARCHITECT / 主创建筑师：Carlos Ott, Silvia Menéndez Associated Architect
ASSOCIATED ARCHITECTS / 合作建筑师：Estudio Hauser, Ziblat
ARCHITECT OF RECORD / 合作方：Ing. Mario Di Marco + MARQ

This is a 12-storey residential corner building with roof top amenities and two retail at street level. Each floor and unit vary with different angular balconies. Parking garages below grade. The crossing diagonal balconies shapes provide a broken facade in one of the city´s old neighbourhood.

这幢转角建筑共有12层，带屋顶设施，两个街面商铺。每个楼层和住宅拥有不同角度的阳台。停车库设在地下。交错的斜角阳台在这座城市的旧街区中显得醒目出挑。

LOS PINOS EDUCATIONAL CENTRE
洛斯皮诺斯教育中心

Educational / 教育

Montevideo, URUGUAY / 蒙得维的亚, 乌拉圭

LOCATION / 地址：San Martín Ave. 5654, Casavalle
CLIENT / 业主：Fundación Los Pinos
AREA / 面积：1800 m^2
HEIGHT / 高度：2 storeys / 2层
UNITS / 户数：13 classrooms / 13间教室
DATE / 设计时间：2016
STATUS / 状态：Under Construction / 建设中
DESIGN ARCHITECT / 主创建筑师：Carlos Ott
ASSOCIATED ARCHITECTS / 合作建筑师：Carlos Ponce de Leon Associated Architects

Los Pinos is an Educational Centre that promotes the integral development of the children of low-income families of Casavalle. The new building is developed on two levels with a total of 13 classrooms for primary and secondary levels, services and administrative premises. Ths building has a concrete structure with metal steel roofing, stone walls and solar panels.

该中心是一个促进卡萨瓦莱低收入家庭子女全面发展的教育中心。新大楼分为两层，共有13间教室，用于中小学教学、服务和行政办公场所。建筑带有金属钢屋顶、石墙和太阳能电池板。

EZ ESTHER TOWERS
EZ埃斯特大厦

Commercial Offices / 商业办公
São Paulo, BRAZIL / 圣保罗, 巴西

LOCATION / 地址：New Dr. Chucri Zaidan Ave. & José Vicente Cavalheiro, Morumbi
CLIENT / 业主：EZ TEC Empreendimentos e Participações S.A.
AREA / 面积：136 500 m^2
HEIGHT / 高度：139 m, 27 storeys / 27层
UNITS / 户数：34 open offices / 34间开放办公室
DATE / 设计时间：2009
STATUS / 状态：Completed 2014 / 2014年建成
DESIGN ARCHITECT / 主创建筑师：Carlos Ott
ARCHITECT OF RECORD / 合作方：Silvio Iamamura

Two symmetrical 27-storey sloping towers, comprised by seventeen open office floors, situated in a São Paulo key trade zone close to malls and main new corporate buildings, distinguished by an inverted pyramidal helipad that suspends on the central axis joining the inclined peaks. A 6-storey slanting parking podium includes an iconic restaurant building that relates the pedestrian commercial promenade. Clad in dark grey, white and silver reflective glass, the inclined elevations pretend to provide a dynamic appearance to this office complex in the best location of the city. A fountain and waterfall at the centre of the vehicular and pedestrian access create a roundabout element giving access under large glass canopies to the two different office towers´ lobbies with a major cafe at the centre. The client requested a building that provided a strong three dimensional logo of the complex.

EZ埃斯特大厦由两座对称的27层倾斜塔楼组成，有17层开放式的办公层。大厦位于圣保罗主要贸易区，靠近购物中心和商务区，其特点是拥有一个倒金字塔形的直升机停机坪，该直升机停机坪悬挂在两座倾斜的塔楼中心。大厦还有一幢6层高的倾斜裙房，包含一座与步行商业长廊相连的标志性的餐厅。覆盖着深灰色、白色和银色的反光玻璃，上升的立面为处于城市最佳位置的办公大楼提供了动感的外观。车辆和行人通道的中心处有喷泉和瀑布，形成了环形交叉路口。大玻璃雨棚下可通向两个不同的办公大厅，中间有一个咖啡馆。根据客户要求，建筑设计成具三维标识的复杂建筑。

JADE PARK
翡翠公园公寓

Residential / 住宅

Asunción, PARAGUAY / 亚松森，巴拉圭

LOCATION / 地址：Santísima Trinidad Ave. & Congreso Colombia Ave.
CLIENT / 业主：Fortune International Group, Jiménez Gaona y Lima S.A., Grupo Barcos y Rodados, Tierra Buena S.A.
AREA / 面积：42 600 m^2
HEIGHT / 高度：143 m
UNITS / 户数：122
DATE / 设计时间：2014
STATUS / 状态：Under construction / 建设中
DESIGN ARCHITECT / 主创建筑师：Carlos Ott
ARCHITECT OF RECORD / 合作方：Jiménez Gaona y Lima S.A.

Prime residential complex including three towers of 4, 3 and 2 bedrooms, a large clubhouse with full amenities is in one of Asuncion top locations and a site with magnificent vegetation, tennis and football courts; and landscaped outdoors.Units have large balconies with gas fueled barbeques and granite and marble finishes. Three Organic towers with curvilinear balconies sitting on triple height lobbies take advantage of the very green property with 360 degree views of the city.Due to the strong afternoon heat, the west facades aligning on Santísima Trinidad Ave. contain the service areas and are built in exposed concrete.A longitudinal park runs along the whole length of the avenue, with access from each end of the large site.

位于亚松森豪宅区的翡翠公园公寓包括三座塔楼，户型有两居室、三居室、四居室，小区内拥有设施齐全的大型会所、茂密的植被、网球场和足球场，以及户外景观。公寓花岗岩和大理石饰面的大阳台设有燃气烧烤架。三座带有曲线阳台的绿色塔楼坐落在三层高的大堂上，充分利用了植被的优势，可欣赏城市360度全景。由于强烈的午后高温，位于三圣一大道一侧的西立面设有设备房，并以混凝土建造。一个纵向的公园沿着大道延伸，可以从公园的两端进入。

PORTOMARINE
波尔托马林公寓

Residential / 住宅

Cartagena, COLOMBIA /卡塔赫纳, 哥伦比亚

LOCATION / 地址：Bahía de Cartagena - Avenida Sucre y Nariño, Bocagrande
CLIENT / 业主：Vía Grupo S.A.S. José Antonio Villegas Vélez
AREA / 面积：18 000 m^2
HEIGHT / 高度：163 m, 40 storeys / 40层
UNITS / 户数：52
DATE / 设计时间：2016
STATUS / 状态：Under construction / 建设中
DESIGN ARCHITECT / 主创建筑师：Carlos Ott
ARCHITECT OF RECORD / 合作方：Rafael Cepeda

In a unique location on the edge of Cartagena de Indias bay, the structure has 40 floors of one or two prime residences per floor on top of an 8-storey garage, and amenities at the top of the parking levels and roof top with 360 views of the city, bay and ocean.The flute appearance of the structure is obtained by the use of brise soleils around its perimeter at each apartment level.The horseshoe plan opens curved terraces toward the bay frontage while maintaining a more opaque facade on the city side.

该建筑坐落在迦太基海湾边缘的一个独特位置，有一个8层车库，在其上有40层住宅，每层有一户或两户单元，在停车楼的顶部和屋顶有设备房，大楼可以360度欣赏城市、海湾和海洋全景。该建筑的长笛形外观是通过在每层公寓周围使用遮阳板获得的。马蹄形平面布局向海湾一侧为圆弧形的露台，而面向城市一侧设计为不太透明的外观。

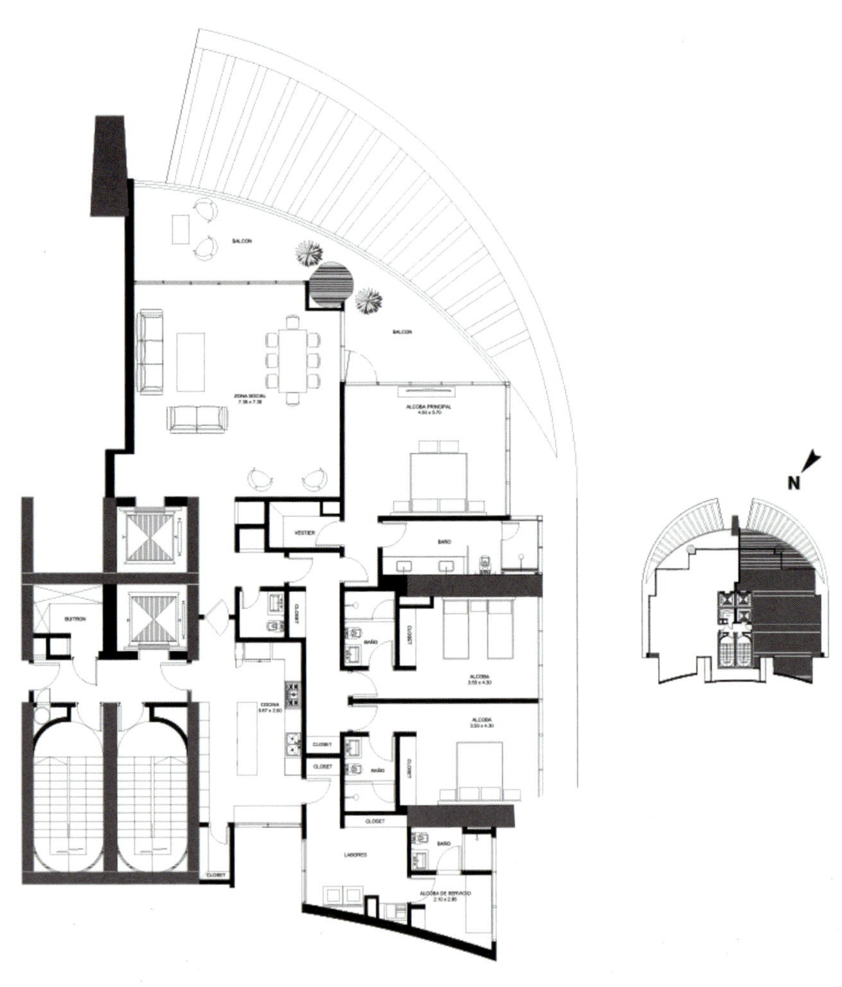

THE HIGHLANDS
高地公寓

Residential / 住宅

North Miami Beach, Florida, USA / 北迈阿密海滩, 佛罗里达, 美国

LOCATION / 地址：13800 Highland Drive, North Miami Beach
CLIENT / 业主：Blue Rd LLC
AREA / 面积：6817 m^2
HEIGHT / 高度：6 storeys / 6层
UNITS / 户数：60
DATE / 设计时间：2015
STATUS / 状态：Completed 2018 / 2018年建成
DESIGN ARCHITECT / 主创建筑师：Carlos Ott
ARCHITECT OF RECORD / 合作方：Frankel Benayoun Architects Inc.

6-storey condominium with 60 units is located in one of the islands in Florida Intracoastal Waterway. The structure provides different units typology on a prime residential neighbourhood.

该建筑有6层，共有60个单元，位于佛罗里达内陆水道的一个岛屿上，在高档住宅区提供了不同的住宅户型。

QUADRO JURAMENTO
夸德罗·胡拉门托公寓

Residential / 住宅

Buenos Aires, ARGENTINA / 布宜诺斯艾利斯, 阿根廷

LOCATION / 地址：4242 Juramento St., Belgrano, Villa Urquiza
CLIENT / 业主：YAPAY Desarrolladores
AREA / 面积：7600 m^2
HEIGHT / 高度：5 storeys / 5层
DATE / 设计时间：2015
STATUS / 状态：Completed 2019 / 2019年建成
DESIGN ARCHITECT / 主创建筑师：Carlos Ott
ARCHITECT OF RECORD / 合作方：José Garat

A curvilinear double-height hall gives entrance to this corner building formed by turning open boxes with protruding slabs and columns making different shapes for each unit. Include two, three and four-bedroom units vary from 44 to 200 m^2 and a basement parking. Amenities include a multipurpose room, a gym, changing rooms, solarium, terraces, sundecks, gardens and a swimming pool.

建筑位于街角，入口为一个曲线形的双层高大厅，该建筑通过外凸楼板和圆柱组合形成犹如叠在一起的盒子，从而使每户住宅的形状各不相同。其中包括两居、三居和四居室（面积从44 m^2到200 m^2不等）和一个地下停车场。配套设施包括多功能室、健身房、更衣室、日光浴室、露台、日光浴平台、花园和游泳池。

UNKANNY
翁卡尼公寓

Residential / 住宅

Mar del Plata, ARGENTINA / 马德普拉塔, 阿根廷

LOCATION / 地址：Leandro N. Alem St. & Gascón St.
CLIENT / 业主：H. Ledesma Desarrollos Inmobiliarios
AREA / 面积：20 000 m^2
HEIGHT / 高度：19 storeys / 19层
UNITS / 户数：93
DATE / 设计时间：2015
STATUS / 状态：Under construction / 建设中
DESIGN ARCHITECT / 主创建筑师：Carlos Ott
ARCHITECT OF RECORD / 合作方：Mariani Perez Maraviglia Cañadas

In the original site of the house of Mariano Mores, a famous Argentinian musician, the high class residential tower respects the existing mansion that is used as a clubhouse. The prime location on the ocean front in a high site gives unimpeded views of Mar del Plata, Argentina´s prime summer resort.

该项目位于阿根廷著名音乐家马里亚诺·莫尔斯的房屋原址上，高层住宅楼与原址现存的公馆相互融合，原公馆现作为俱乐部。大楼位于海滨地势较高处，可以一览无余地欣赏阿根廷最主要的避暑胜地马德普拉塔的景色。

HILTON HOTEL AND AMAMBAY BANK CENTRE 希尔顿酒店和阿曼贝银行中心

Mixed use / 综合使用

Asunción, PARAGUAY / 亚松森，巴拉圭

LOCATION / 地址：Aviadores del Chaco Ave. & H. Campos Cervera St.

CLIENT / 业主：Developy S.A.

AREA / 面积：83 000 m^2

HEIGHT / 高度：114 m, 24 storeys / 24层

UNITS / 户数：180 rooms, 70 apartments / 180间客房，70间住宅

DATE / 设计时间：2015

STATUS / 状态：Under construction / 建设中

DESIGN ARCHITECT / 主创建筑师：Carlos Ott

ARCHITECT OF RECORD / 合作方：Jiménez Gaona y Lima S.A.

The project is a multi-use complex including a 180-room hotel and 70 serviced apartments, a private bank building, a multipurpose building housing a newspaper, television channel and publicity agency as well as a below grade parking garage. Located in Asuncion main thoroughfare, the buildings are clustered around an open plaza with retail and restaurants.The concept was to design different volumetric forms clad in different type of glass with the purpose of providing a variety of forms but with a restrained palette.The project creates a major outdoor pedestrian public space that is expected to have lots of activities as being surrounded by retail, in probably Asuncion´s most important commercial avenue.

该项目是一个多用途综合大楼，包括180间客房的酒店和70间住宅的服务型公寓，一栋私人银行大楼，一栋容纳报纸、电视频道和宣传机构的多功能大楼，以及一个地下停车场。建筑群位于亚松森的主要道路，周围是一个散布着零售商店和餐厅的开放广场。其概念是设计包覆在不同类型玻璃幕墙之下的不同楼宇，以提供各种形式但相似颜色的建筑群。该项目在亚松森最重要的一个商业街区创建了拥有众多户外活动和零售空间的公共步行场所。

Hilton

ALLURE
阿鲁尔公寓

Residential / 住宅

Cartagena, COLOMBIA / 卡塔赫纳, 哥伦比亚

LOCATION / 地址：Carrera 5ta & Carrera 8va, Bocagrande
CLIENT / 业主：KMA Construcciones
AREA / 面积：26 000 m^2
HEIGHT / 高度：182 m, 40 storeys / 40层
UNITS / 户数：41
DATE / 设计时间：2016
STATUS / 状态：Under construction / 建设中
DESIGN ARCHITECT / 主创建筑师：Carlos Ott
ARCHITECT OF RECORD / 合作方：Santiago Barriga Fayad

Situated in a prime location with direct views and access to Cartagena bay, the 40-storey luxury residential building has two large units in the typical floors, duplex apartments at the top, and sits on an eight storeys above grade parking garage. Full amenities are located at the top of the garage and a helipad is situated at the top of the structure.

这栋40层高的豪华住宅坐落在享有最优景观并可以直达卡塔赫纳海湾的黄金地段，大楼包含两个部分，顶部的复式公寓以及底部的8层停车楼。完备的配套设施房位于车库的顶部，大楼的顶部建有停机坪。

ALLURE

NOSTRUM BAY
诺斯特姆湾公寓

Residential, Commercial / 住宅, 商业
Montevideo, URUGUAY / 蒙得维的亚, 乌拉圭

LOCATION / 地址：Rambla Franklin D. Roosevelt & Rambla Sud America, Aguada
CLIENT / 业主：Altius Group
AREA / 面积：19 089 m²
HEIGHT / 高度：22 storey / 22层
UNITS / 户数：194
DATE / 设计时间：2018
STATUS / 状态：Under construction / 建设中
DESIGN ARCHITECT / 主创建筑师：Carlos Ott
ARCHITECT OF RECORD / 合作方：Carlos Ponce de León Architects

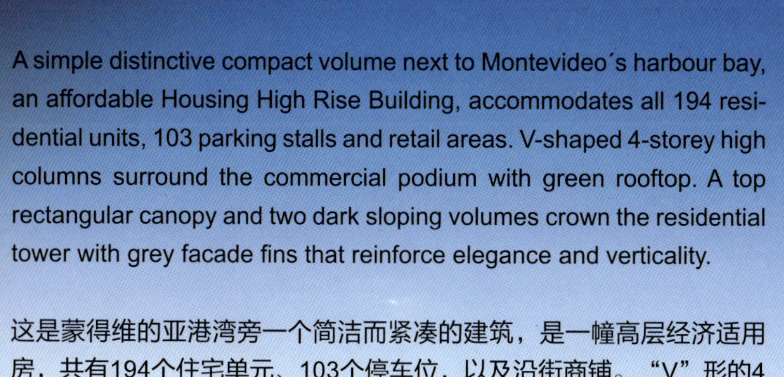

A simple distinctive compact volume next to Montevideo´s harbour bay, an affordable Housing High Rise Building, accommodates all 194 residential units, 103 parking stalls and retail areas. V-shaped 4-storey high columns surround the commercial podium with green rooftop. A top rectangular canopy and two dark sloping volumes crown the residential tower with grey facade fins that reinforce elegance and verticality.

这是蒙得维的亚港湾旁一个简洁而紧凑的建筑，是一幢高层经济适用房，共有194个住宅单元、103个停车位，以及沿街商铺。“V”形的4层高圆柱围绕着拥有绿色屋顶的商业裙房。顶部为矩形顶篷和两个深色的倾斜体块，建筑表面为灰色立面，增强了优雅和垂直感。

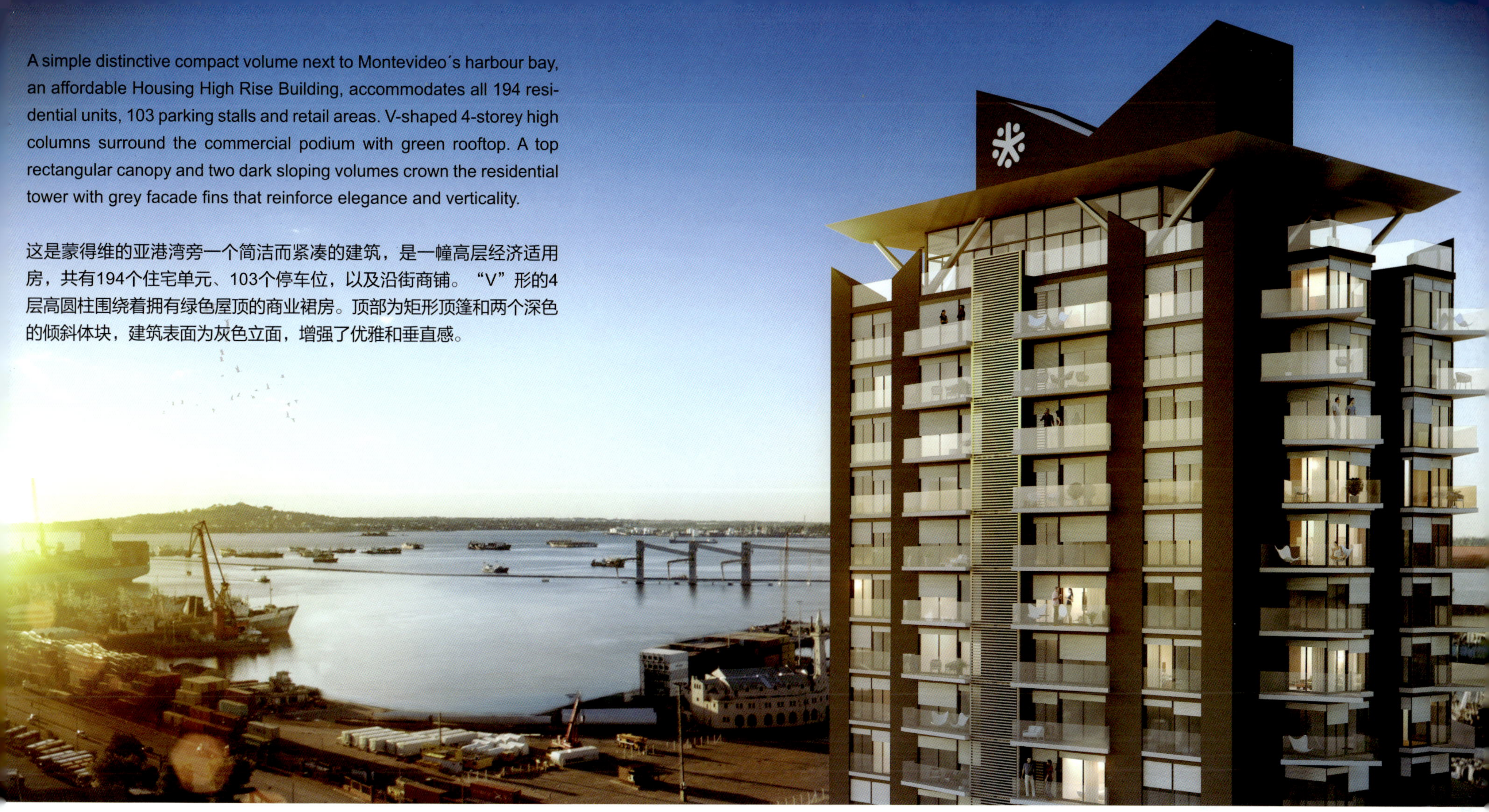

ALMA CORSO
阿尔玛科索公寓

Residential / 住宅

Montevideo, URUGUAY / 蒙得维的亚, 乌拉圭

LOCATION / 地址：Canelones & Convencion St., Downtown Montevideo

CLIENT / 业主：Mikeldir S.A.

AREA / 面积：5600 m^2

HEIGHT / 高度：27 m, 12 storeys / 12层

UNITS / 户数：57

DATE / 设计时间：2016

STATUS / 状态：Under construction / 建设中

DESIGN ARCHITECT / 主创建筑师：Carlos Ott, Silvia Menéndez Architect

ASSOCIATED ARCHITECT / 合作建筑师：German Hauser

ARCHITECT OF RECORD / 合作方：Ing. Mario Di Marco + MARQ

Twelve storeys residential building contains outdoor barbeques and amenities at the top and retail at street level in the Montevideo downtown area. In order to create a different image, balconies are housed in glass blocks with inclined windows.

该项目共有12层，顶部设有空中烧烤设施，临街有商铺。为了创建不同的形象，阳台被设计成带有倾斜窗户的玻璃体。

DELI

IVENS DIAS BRANCO
伊文斯迪亚斯布朗克大厦

Residential / 住宅

Fortaleza, Ceará, BRAZIL / 福塔莱萨, 塞阿拉, 巴西

LOCATION / 地址：Beira Mar Ave. & Avenida Barão de Sudart Ave., Meireles
CLIENT / 业主：Idibra Participações
AREA / 面积：40 000 m^2
HEIGHT / 高度：139 m, 40 storeys / 40层
UNITS / 户数：31
DATE / 设计时间：2016
STATUS / 状态：Under construction / 建设中
DESIGN ARCHITECT / 主创建筑师：Carlos Ott
ARCHITECT OF RECORD / 合作方：CIA de Arquitetura-Osvaldo Luis Freitas de Souza

Located in a prime site over the Atlantic Ocean beach of Fortaleza city in north east Brazil, the 40-storey building has 600 m^2 large units per floor including a helipad at its top and large terraces around its perimeter. In order to accentuate its location on the edge of spectacular beaches, the building is conceived as a sinuous sculpture mimicking the ocean waves at its feet, with curvilinear deep balconies around its three waterfront sides.The rear facade is on the contrary facing the city is treated as a more opaque wall locating all services in its perimeter.

这座40层高的建筑位于巴西东北部福塔雷萨市大西洋海滩上的黄金地段，每层楼拥有600 m^2的住宅单元，建筑包括顶部的停机坪和其周围的大露台。为了突出其在壮观的海滩边缘的位置，设计师模仿海浪，将建筑设计为曲线形的雕塑，并在其3个滨水面设计了弧形大阳台。相反，面向城市的立面被设计为不透明的墙，所有服务功能设施都设置在这一侧。

QUADRO HOLMBERG
夸德罗·霍姆伯格公寓

Residential / 住宅

Buenos Aires, ARGENTINA / 布宜诺斯艾利斯, 阿根廷

LOCATION / 地址：Holmberg St. & Blanco Encalada St., Belgrano
CLIENT / 业主：YAPAY Desarrolladores
AREA / 面积：8830 m^2
HEIGHT / 高度：5 storeys / 5层
UNITS / 户数：51
DATE / 设计时间：2017
STATUS / 状态：Under construction / 建设中
DESIGN ARCHITECT / 主创建筑师：Carlos Ott
ARCHITECT OF RECORD / 合作方：Lopatín Architects

It is a 5-storey residential building located in the New Juramento Street corridor of Buenos Aires. The building has different typology of units including duplex units, studios, one, two and three bedrooms apartments and penthouses. Parking garage is below grade.Using exposed concrete prefabricated elements, the curved balconies provide a varied frontage to the building clearly identifying each individual unit from the outside.A very transparent entrance in the corner houses the main lobby of the building.

这是一幢5层高的住宅楼，位于布宜诺斯艾利斯的新茹拉门图大街之上，具有不同的单元类型，包括复式单元、工作室、一居室、两居室和三居室公寓和顶层公寓，以及地下停车库。弧形阳台使用裸露的混凝土构件，为建筑物提供了多样化的正面效果，从外部可清晰地识别出每个单独的单元。拐角处的大厅非常透明，是建筑物的主要入口。

FISHERTON PARK
菲舍顿公园公寓

Residential / 住宅

Rosario, ARGENTINA / 罗萨里奥, 阿根廷

LOCATION / 地址：Brassey St. & Terragona Rd., Fisherton
CLIENT / 业主：Dix Desarrollos Urbanisticos
AREA / 面积：10 400 m^2
HEIGHT / 高度：10 m, 3 storeys / 3层
UNITS / 户数：42
DATE / 设计时间：2018
STATUS / 状态：Under construction / 建设中
DESIGN ARCHITECT / 主创建筑师：Carlos Ott
ARCHITECT OF RECORD / 合作方：Mascetti Lavarello Arquitectos

Located in a heavily landscaped community in the city, with centennial trees, the low height project is situated in the centre of the site respecting the surrounding vegetation. It is provided with large terraces and roof gardens to different units that are through tipe with access from the street frontage and opening to the centre of the site where the original residence is adapted to become the clubhouse of the compound with amenities and swimming pool.

项目位于城市一个拥有百年树木的深度植被社区，设计时降低了建筑整体高度以适应周围的植被，并为不同的单元提供了大型露台和屋顶花园，这些露台可以从临街通道进入，并向社区中心开放，原来的住宅被改造成带有服务设施和游泳池的小区会所。

CAPRI RESIDENCES
卡普里公寓

Residential / 住宅

North Miami Beach, Florida, USA / 北迈阿密海滩, 佛罗里达, 美国

LOCATION / 地址：Highlands Drive
CLIENT / 业主：Blue Rd Developers LLC
AREA / 面积：46 800 m^2
HEIGHT / 高度：15 storeys / 15层
UNITS / 户数：257
DATE / 设计时间：2018
STATUS / 状态：Under construction / 建设中
DESIGN ARCHITECT / 主创建筑师：Carlos Ott
ARCHITECT OF RECORD / 合作方：Frankel Benayoun Architects

It is a residential building with apartments and townhouses in the proximity of Biscayne Boulevard, one of Miami's most important streets. It has full amenities and penthouses with private pools.A variety of residential types are provided with access to a major outdoor terrace that includes a large open swimming pool.Penthouses with private pools are also provided at the roof top level.

比斯坎大道是迈阿密最重要的街道之一，这个带有公寓和联排别墅的住宅楼就位于比斯坎大道附近，并配备齐全的设施和带私人泳池的顶层公寓。该公寓提供各种类型的住宅，可使用包括大型开放式游泳池的主要室外露台。顶层还设有带私人游泳池的顶层公寓。

PUNTA DEL ESTE TOWER
埃斯特角公寓

Residential / 住宅

Cartagena, COLOMBIA / 卡塔赫纳, 哥伦比亚

LOCATION / 地址：Calle 5a, Castillogrande

CLIENT / 业主：Pezzano & Saieh S.A.S.

AREA / 面积：5600 m²

HEIGHT / 高度：74 m, 16 storeys / 16层

UNITS / 户数：34

DATE / 设计时间：2018

STATUS / 状态：Under construction / 建设中

DESIGN ARCHITECT / 主创建筑师：Carlos Ott

ARCHITECT OF RECORD / 合作方：Ivan Turriago Thorrens

Located in Castillogrande, Cartagena, with sea and bay views, the tower formed by a set of volumes, canopies and terraces that alternate creating a rhythmic image of dynamism. Flanking a central vertical volume with louvers, balconies vary in size and location breaking the strict verticality usually seen in a tower. Luxurious amenities on the roof top, crown the 16 storeys of two apartments per floor, the lower playground amenities and the double height lobby.

该建筑位于卡塔赫纳的卡斯提洛格兰德，可欣赏大海和海湾的景色。公寓由一组建筑体、顶棚和露台组成，交替创造出充满活力的韵律形态。与通常的塔楼不同，该建筑带有遮阳板的中央垂直体块的两侧，阳台的大小和位置各不相同。建筑共16层，每层有两套公寓，低层设有游乐场所和两层高的大堂，顶层设有豪华设施。

MACA MUSEO DE ARTE CONTEMPORÁNEO AMERICANO
美洲当代艺术博物馆

Museum / 博物馆

Punta del Este, Maldonado, URUGUAY/ 埃斯特角, 马尔多纳多, 乌拉圭

LOCATION / 地址：Ruta 104, Km 4.500, Manantiales

CLIENT / 业主：Fundación Pablo Atchugarry

AREA / 面积：4050 m^2

HEIGHT / 高度：15.75 m

DATE / 设计时间：2018

STATUS / 状态：Under Construction / 建设中

DESIGN ARCHITECT / 主创建筑师：Carlos Ott

ARCHITECT OF RECORD / 合作方：Estudio de Arquitectura Atchugarry

The Pablo Atchugarry Foundation, a non-profit institution created by the sculptor to promote Arts situated in a country complex in Manantiales, Punta del Este where will nest the museum. A triangular warped surface covers the glassy connection with the existing Exhibition Hall.

A spacious naturally lit, fully glazed curvilinear casing of laminated wood structure with metal ceilings houses the sculptures. An intimate, introvert concrete and brick prism, with control over UV light, temperature and humidity, encloses the paintings space. The sharp exposed foundation is a reinforced concrete sculpture in itself.

巴勃罗·阿特楚加利基金会是由雕塑家巴勃罗·阿特楚加利（Pablo Atchugarry）创建的一个非营利机构，旨在促进位于埃斯特角马南蒂亚莱斯的一个乡村综合体的艺术氛围，该博物馆就位于此处。一条连接现有展览厅的玻璃通道被一个三角形建筑体覆盖。

全玻璃的弧形层压木结构外壳和金属天花板所构成的宽敞空间充分利用了自然光，里面放置了展览的雕塑作品。绘画作品展示厅由混凝土和砖棱柱围成，可调节紫外线、温度和湿度。暴露出来的锐利地基本身就是一个钢筋混凝土雕塑。

LIVE BELGRANO
贝尔格拉诺公寓

Residential / 住宅

Belgrano, Buenos Aires, ARGENTINA / 贝尔格拉诺，布宜诺斯艾利斯，阿根廷

LOCATION / 地址：2690 Teodoro García & Moldes St.
CLIENT / 业主：Grupo Massana
AREA / 面积：5300 m^2
HEIGHT / 高度：40 m, 12 storeys / 12层
UNITS / 户数：48
DATE / 设计时间：2020
STATUS / 状态：Under construction / 建设中
DESIGN ARCHITECT / 主创建筑师 ：Carlos Ott

2690

On a corner site located in Belgrano, Buenos Aires, Teodoro Garcia, a 12-storey residential building, exposes a fully glazed corner and solid trapeziums framing balconies over both side streets. The ground floor remarks the access lobby with a sculpture in the corner.

在布宜诺斯艾利斯贝尔格拉诺的一个拐角处，一栋12层高的住宅楼向街道两侧展现了全玻璃的转角和坚固的梯形阳台。底层拐角处有一个雕塑，标志着出入大厅。

AIR BROOKLIN
布鲁克林空中公寓

Commercial, Residential / 商业, 住宅
São Paulo, BRAZIL / 圣保罗, 巴西

LOCATION / 地址：4800 Santo Amaro Ave., Brooklin
CLIENT / 业主：EZ TEC Empreendimentos e Participações S.A.
AREA / 面积：70 600 m²
HEIGHT / 高度：115 m
UNITS / 户数：588 apartments, 16 offices / 588间公寓，16间办公室
DATE / 设计时间：2015
STATUS / 状态：Under construction / 建设中
DESIGN ARCHITECT / 主创建筑师：Carlos Ott
ARCHITECT OF RECORD / 合作方：Silvio Iamamura

A major complex including an office building and large residential tower over a parking podium with amenities over the podium and the roof overlooking the city of São Paulo. The developer goal was to provide unique amenities at the top of the tall structure to all residents, thus a major roof garden and park was designed including swimming pools, recreational and gym facilities.The ondulating facade obtained by the use of curved balconies is carried along in the louvres on the parking level. A second green park is located on top of the podium.An office building is built in close proximity providing all possible uses for residents.

这个大型综合体位于圣保罗，包括办公楼和位于停车平台上的大型住宅楼，平台上设有配套设施，屋顶可俯瞰圣保罗市。开发商的目标是在高层建筑的顶部为所有居民提供独特的配套设施，因此设计了一个主要的屋顶花园和公园，包括游泳池、娱乐和健身设施。通过使用弯曲的阳台使立面显得起伏动感，这一点在停车层的百叶窗中得以延续。另一个绿色公园则位于裙房的顶部。附近建造了一座办公楼，为居民提供了所有可能用到的设施。

EZ INFINITY
EZ 无界公寓

Commercial, Residential / 商业, 住宅
São Paulo, BRAZIL / 圣保罗, 巴西

LOCATION / 地址：23 de Maio Ave. & Aquiles Masetti St.
CLIENT / 业主：EZ TEC Empreendimentos e Participações S.A.
AREA / 面积：64 000 m^2
HEIGHT / 高度：82.7 m, 24 storeys / 24层
UNITS / 户数：88
DATE / 设计时间：2020
STATUS / 状态：Under design / 建设中
DESIGN ARCHITECT / 主创建筑师：Carlos Ott
ARCHITECT OF RECORD / 合作方：Silvio Iamamura

Two residential 24-storey towers forming a right angle, located on the corner of 23 de Maio Ave. and Masetti St., contain 5 basement parking levels, a double-height lobby, restaurant and amenities, 21 residential storeys and 2 duplex storeys. Each tower portrays a central core flanked by 3 and 4-bedroom apartments, with external corners surrounded by curved balconies.

两座24层高的住宅楼形成直角。住宅包含5个地下停车场，有双层高大堂、餐厅和配套设施，21个住宅层和两个复式楼层。每座塔楼有中央核心筒，两侧是三居室和四居室公寓，外部角落被弯曲的阳台所包围。

TOYOTOSHI HEADQUARTERS
丰田总部

Commercial, Retail / 商业, 零售
Asunción, PARAGUAY / 亚松森, 巴拉圭

LOCATION / 地址：Mariscal López Ave. & Reclus St.
CLIENT / 业主：Toyotoshi S.A.
AREA / 面积：16 000 m^2
HEIGHT / 高度：47 m, 8 storeys / 8层
DATE / 设计时间：2017
STATUS / 状态：Under design / 建设中
DESIGN ARCHITECT / 主创建筑师：Carlos Ott
ARCHITECT OF RECORD / 合作方：Fabrizio Bibolini

The Toyotoshi Headquarters located in the main commercial avenue of Asuncion, occupies a full city block, the complex includes a major automobile showroom and garage and a 9 storey office building. A major automobile group is housed in these facilities. In addition to a whole block showroom, a below grade service garage, it will incorporate the group headquarters.Due to the many social activities of the owner, the showroom doubles as a large gathering place for major social events.For the benefit of employees, restaurant, library, meeting rooms and a gymnasium as well as a running track are provided in the premises.

这座建筑位于亚松森的主要商业大道，占据了一整个城市街区。该综合体包括一个主要的汽车陈列室，车库以及一座9层高的办公楼。一个主要的汽车集团将被安置于此建筑中，除了占据整个街区的汽车陈列室和地下服务车库外，集团总部也会入驻于此。由于业主的许多社交活动，陈列室兼作重大社交活动的大型聚会场所。为了方便员工，总部设有餐厅、图书馆、会议室和健身房以及跑道。

esplendor
A WYNDHAM GRAND HOTEL

ESPLENDOR PUNTA CARRETAS HOTEL
普兰多庞塔卡雷塔斯酒店

Hotel / 酒店

Montevideo, URUGUAY / 蒙得维的亚, 乌拉圭

LOCATION / 地址：Bulevar Artigas Ave. and Manuel Errazquin, Punta Carretas
CLIENT / 业主：Traripel S.A. - Emprenurban Desarrollos
AREA / 面积：19 657 m^2
HEIGHT / 高度：38 m, 13 storeys / 13层
UNITS / 户数：279
DATE / 设计时间：2009
STATUS / 状态：Completed 2017 / 2017年建成
DESIGN ARCHITECT / 主创建筑师：Carlos Ott, Silvia Menéndez Associated Architect
ARCHITECT OF RECORD / 合作方：Walter Varela Eng.

Esplendor by Wyndham Montevideo is a 4-star hotel comprised by 2 blocks. Over Bulevar Ave., a building of 13 storeys and Roof with Lounge Bar, view towards Montevideo boulevard. Over Errazquin St., access to two commercial premises and a bar, block GF + 6, Solarium, Outdoor Rooftop Pool. Both blocks linked by a covered patio containing the reception and the elevators to the rooms and the public elevators to the basement. Stairs and elevators link to first basement with spa, banquet hall and two additional underground levels of garages, technical areas and warehouses containing 108 parking stalls. The exterior facades are clad in aluminum carpentry with DVH with thermal protection, and decorative vertical and horizontal perforated aluminium louvers on both facades. The frontage to streets portray decks for gastronomic use, planters, and landscape. The ground floor, linked as a commercial, gastronomic promenade, contains lobby lounge spaces illuminated from both facades and by the skylight over the central courtyard.

这个四星级酒店位于街区转角，其中一部分临布勒瓦尔大街，屋顶带有酒吧，并能俯瞰整个蒙得维的亚大道。另一部分临伊斯昆大街，这一侧的入口可以抵达酒店的两个商业区域以及酒吧、日光浴室及屋顶泳池。酒店的两个部分由一个带顶棚的露台相连，露台包含接待处和通往房间的电梯以及通往地下室的公共电梯。楼梯和电梯与水疗中心、宴会厅和地下两层车库，技术区和包含108个停车位的仓库相连。外立面采用DVH铝包覆，具有热保护作用，装饰性、垂直和水平的铝百叶窗包覆于两个立面上。开向街道的立面用于餐饮、景观。一楼与商业美食长廊相连，有大堂休息空间。中央庭院上方为天窗。

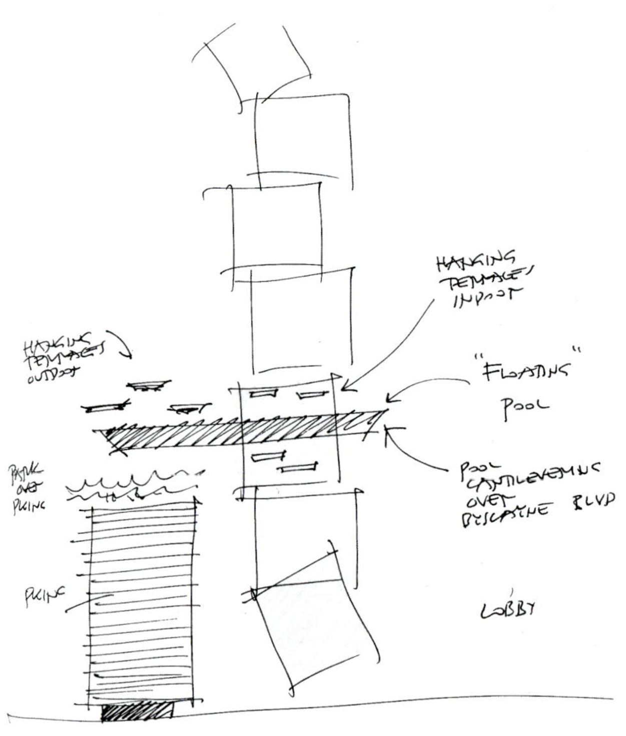

WALDOLF ASTORIA RESIDENCES MIAMI
迈阿密华尔道夫酒店综合体

Residential & Hotel / 住宅，酒店

Downtown Miami, Florida, USA / 迈阿密市中心, 佛罗里达, 美国

LOCATION / 地址：300-330 Biscayne Blvd.
CLIENT / 业主：Property Market Group
AREA / 面积：294 453 m^2
HEIGHT / 高度：320 m, 100 storeys / 100层
UNITS / 户数：360 private residences, 205 hotel guest rooms / 360间公寓，205间酒店客房
DATE / 设计时间：2016 – 2020
STATUS / 状态：Under construction / 建设中
DESIGN ARCHITECT / 主创建筑师：Carlos Ott
ARCHITECT OF RECORD / 合作方：Sieger Suarez Architects

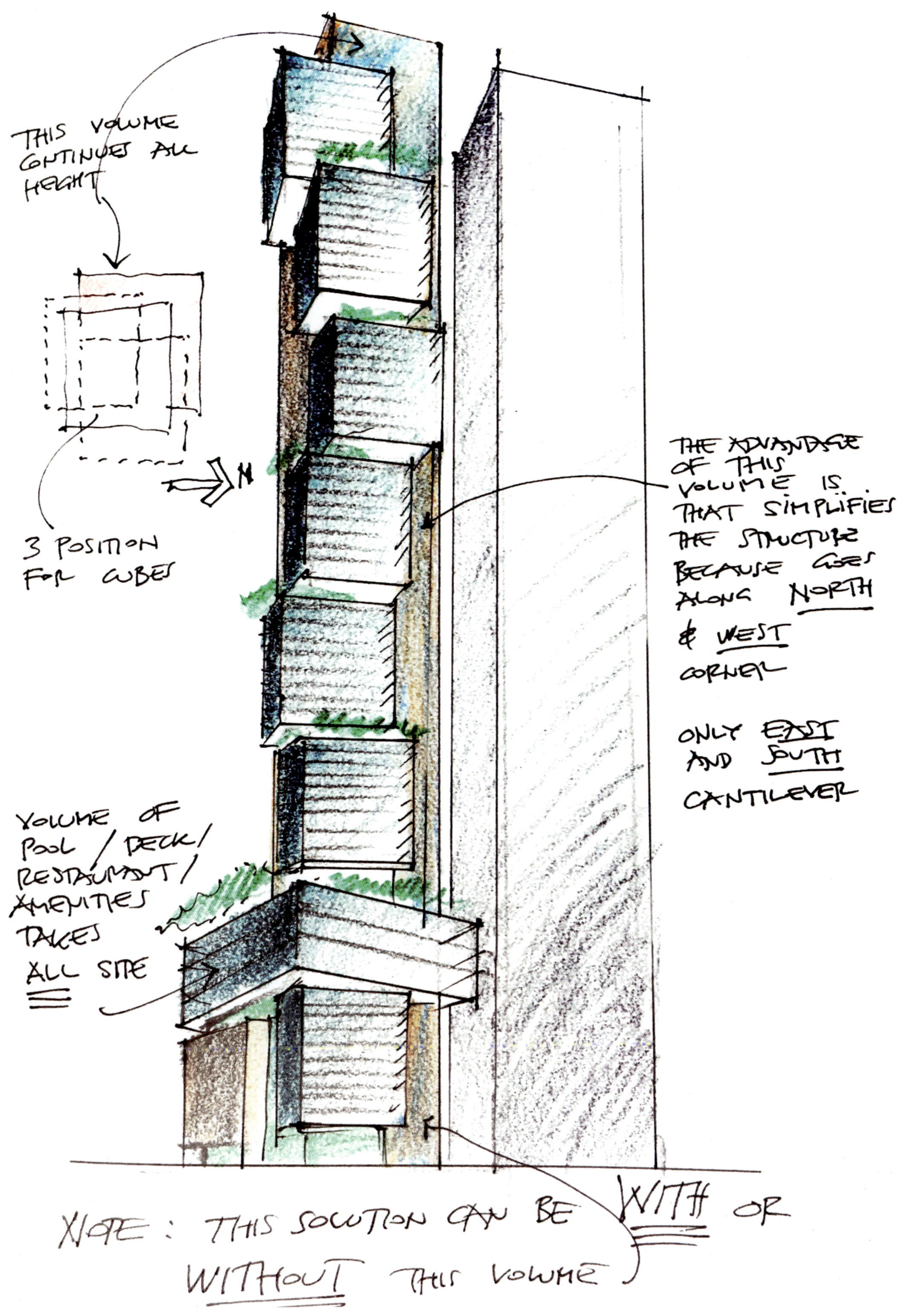
THIS VOLUME CONTINUES ALL HEIGHT
3 POSITION FOR CUBES
N
THE ADVANTAGE OF THIS VOLUME IS THAT SIMPLIFIES THE STRUCTURE BECAUSE GOES ALONG NORTH & WEST CORNER
ONLY EAST AND SOUTH CANTILEVER
VOLUME OF POOL / DECK / RESTAURANT / AMENITIES TAKES ALL SITE
NOTE: THIS SOLUTION CAN BE WITH OR WITHOUT THIS VOLUME

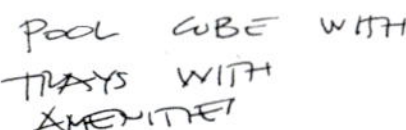

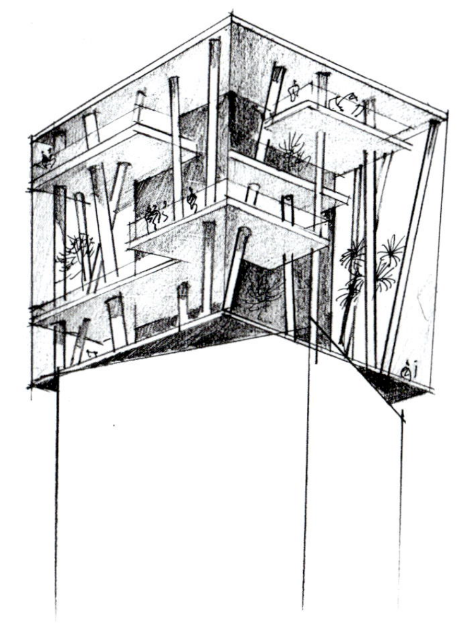

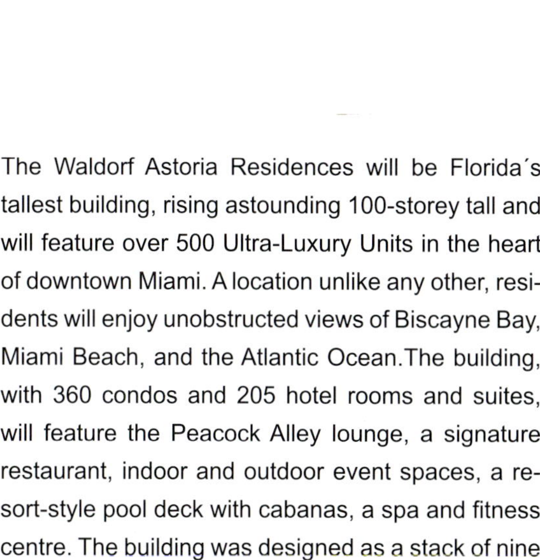

The Waldorf Astoria Residences will be Florida´s tallest building, rising astounding 100-storey tall and will feature over 500 Ultra-Luxury Units in the heart of downtown Miami. A location unlike any other, residents will enjoy unobstructed views of Biscayne Bay, Miami Beach, and the Atlantic Ocean.The building, with 360 condos and 205 hotel rooms and suites, will feature the Peacock Alley lounge, a signature restaurant, indoor and outdoor event spaces, a resort-style pool deck with cabanas, a spa and fitness centre. The building was designed as a stack of nine offset glass cubes creating a moving silhouette as the landmark of downtown Miami.

华尔道夫酒店综合体将成为佛罗里达州最高的建筑，它位于迈阿密市中心，高达100层，拥有500多套超豪华公寓。建筑坐落于一个与众不同的地方，居民将在此享受到比斯坎湾、迈阿密海滩和大西洋一览无余的美景。该建筑拥有360间公寓和205间酒店客房，公寓设有孔雀巷酒廊、特色餐厅、室内和室外活动空间、带泳池的度假小屋、水疗中心和健身中心。该建筑被设计为一系列偏移的玻璃立方体，为迈阿密市中心创造了一个偏移的剪影地标。